Vedran Glavas

Micromechanical Modeling and Simulation of Forming Processes

Schriftenreihe
Kontinuumsmechanik im Maschinenbau
Band 8

Karlsruher Institut für Technologie (KIT)
Institut für Technische Mechanik
Bereich Kontinuumsmechanik

Hrsg. Prof. Dr.-Ing. habil. Thomas Böhlke

Eine Übersicht aller bisher in dieser Schriftenreihe erschienenen Bände finden Sie am Ende des Buchs.

Micromechanical Modeling and Simulation of Forming Processes

by
Vedran Glavas

KIT Scientific Publishing

Dissertation, Karlsruher Institut für Technologie (KIT)
Fakultät für Maschinenbau
Tag der mündlichen Prüfung: 26. Juli 2016

Impressum

KIT Scientific Publishing

Karlsruher Institut für Technologie (KIT)
KIT Scientific Publishing
Straße am Forum 2
D-76131 Karlsruhe

KIT Scientific Publishing is a registered trademark of Karlsruhe Institute of Technology. Reprint using the book cover is not allowed.

www.ksp.kit.edu

Print on Demand 2017

ISSN 2192-693X
ISBN 978-3-7315-0602-7
DOI 10.5445/KSP/1000061958

Micromechanical Modeling and Simulation of Forming Processes

Zur Erlangung des akademischen Grades

Doktor der Ingenieurwissenschaften

der Fakultät für Maschinenbau

Karlsruher Institut für Technologie (KIT)

genehmigte

Dissertation

von

Dipl.-Ing. Vedran Glavas

Tag der mündlichen Prüfung:	26. Juli 2016
Hauptreferent:	Prof. Dr.-Ing. habil. Thomas Böhlke
Korreferent:	Prof. Dr.-Ing. habil. Volker Schulze

Zusammenfassung

In industriellen Anwendungen wird bei Umformprozessen die Texturentwicklung gewöhnlicherweise fast ausschließlich phänomenologisch simuliert. Dies impliziert, dass die Mikrostrukturentwicklung und die daraus resultierende verformungsinduzierte plastische Anisotropie nicht vorhergesagt werden können. Diese induzierte Anisotropie ist aber für die Auslegung und Optimierung industrieller Formgebungsprozesse von großer Bedeutung. Aus diesem Grund wurde im Rahmen dieser Arbeit ein mikromechanisches Materialmodell für die Anwendung bei praxisnahen Umformprozessen entwickelt. Darin wird ein kristallplastisches Materialmodell mit einem nichtlinearen Homogenisierungsansatz kombiniert. Um die Simulation industriell relevanter Probleme zu ermöglichen, wurde das Materialmodell mit Homogenisierungsschema in ein kommerzielles, explizites Finite Elemente Programm auf Gauss Punkt Ebene implementiert. Das verwendete Homogenisierungsverfahren vom Hashin-Shtrikman Typ wurde für den Sonderfall elastisch isotropen und plastisch inkompressiblen Materialverhaltens vereinfacht, weswegen die Verteilung der Spannungs- und Dehnungsfluktuationen von nur einem skalaren Parameter abhängt. Der Minimal- und der Maximalwert des Parameters ergeben als Spezialfall dieser Theorie die klassischen Mischungstheorien von Taylor und Sachs.
Als industrienahe Anwendung wurde der Herstellungsprozess von Getränkedosen aus Aluminium ausgewählt. Für diesen Prozess ist bekannt, dass phänomenologische Modelle nur sehr limitierte Vorhersagen für die Zipfelprofile der Dosen liefern und somit immer noch beträchtliche Einsparpotentiale bei Material und Energie nicht erschlossen werden konnten.

Die Ergebnisse aus den Blechumformsimulationen werden sowohl mit experimentellen Ergebnissen als auch Simulationen eines phänomenologischen Modells verglichen.

Summary

The texture evolution in industrial forming processes is usually simulated with phenomenological material models. Thereby, the evolution of the microstructure and the resulting plastic anisotropy induced by the deformation cannot be predicted. This kind of plastic anisotropy is of great importance for the prediction and the improvement of forming processes in industrial applications. Therefore, a micromechanical model is proposed which combines a crystal plasticity constitutive law with a non-linear homogenization scheme.

The constitutive law with homogenization scheme is implemented in a commercial, explicit finite element code. The implementation at the Gauss point level guarantees computational efficiency in order to perform industrially relevant forming precesses. For the special case of elastic isotropy and plastic incompressibility, a Hashin-Shtrikman type homogenization scheme is developed, with the distribution of stress and strain fluctuations dependent only on a scalar parameter. The minimum and maximum values of the homogenization parameter give the classical mixture theory results of Taylor and Sachs as a special case.

As the industrial application, the aluminum beverage can forming process is chosen. It is well known for this application that phenomenological material models are limited in their prediction of the non-uniform cup height (earing profile). Thus, by predicting the earing profile with higher accuracy there is tremendous potential to reduce costs by saving material and energy consumption. The results of the sheet metal forming simulations are compared to experimental measurements as well as simulations with a phenomenological constitutive law.

Acknowledgments

First and foremost, I would like the express my sincerest gratitude to my advisor Prof. Dr.-Ing. Thomas Böhlke for the continuous support during this thesis. It was his lecture on Engineering Mechanics which inspired me to pursue this field and I am very grateful for the opportunity to work under his guidance.
Further, I would like to thank Prof. Dr.-Ing. Volker Schulze who reviewed this work and served as co-advisor on the committee.
The support of the German Research Foundation (DFG) within the Research Training Group GRK 1483 in the transfer project with the industrial partner Constellium ist gratefully acknowledged. Many thanks to Dr. Dominique Daniel and Dr.-Ing. Chrisian Leppin for their expertise and advice. I benefited very much from the fruitful discussions. All experiments shown in this work have been performed by Connstellium within the transfer project T-A2 of Research Training Group GRK 1483: thank you to Dr.-Ing. Christian Leppin (tensile tests, hydraulic bulge tests, Lankford coefficients, Nakajima tests) and Philippe Bridot (texture measurements and earing profile measurements).
To my colleagues, thank you for creating a pleasant working enviroment, the stimulating talks (on topic - and beyond) and the camaraderie during our time in Karlsruhe. Special thanks to Katja, Barthel and Felix for their guidance and sharing of experience; as well as to Rudolf and Florian for the great cooperation within the GRK 1483. Additionally, thank you to my co-workers Rumena, Stephan, Malte and Eric with whom I shared office space at some point during my stay.

Last but certainly not least, heartfelt thank you to my parents and my sister, without their love, patience and support none of this would have been possible.

Contents

Chapter 1

Introduction

1.1 The can forming process

Cans are widely used in the packaging industry to contain and transport a large variaty of beverages consumed in today's society. The number of annually produced beverage cans amounts to about 250 million of which 90% are made purely from aluminum. Aluminum cans are a lightweight product and are recyclable without a loss in quality. This is a benefit compared to other used packaging materials, such as, for example, plastic-based materials.

Beverage cans are produced in several manufacturing steps including deep drawing, redrawing, ironing, necking and capping. In the first step, a circular blank is cut out of the aluminum coil. Then, the blank is put on a die and fixed with a blank-holder. After that, the cup is formed by deep drawing during the punch motion, see Fig. 1.1.

In the subsequent redrawing step, a similar deep drawing operation is performed using a smaller die and punch. Thereby, the cup height is further increased. Then, three ironing steps are performed with the same punch geometry as used in the redrawing step. With the stepwise decreasing circumference of the die, the wall of the cup is thinned and the cup height increases until the product height is reached.

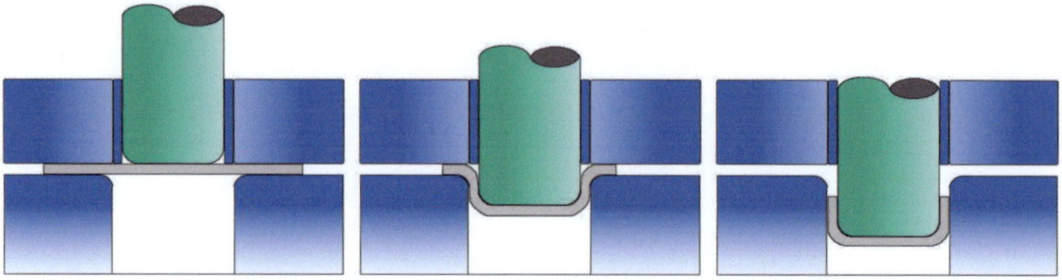

Figure 1.1: Deep drawing of initially circular blank

Predicting the final cup shape accurately is an important task. Even through the reduction of small amount of scrap for every cup there is potential to save costs due to the aforementioned large number of cups produced per year.

1.2 Objective of this work

Phenomenological polycrystal models for sheet metal forming simulations, used in the industrial context, do not produce fully satisfying results. Usually, a lot of material parameters are needed and the parameter identification has been proven to be quite challenging. Therefore, the microstructure must be taken into account in order to develop a more reliable prediction scheme.

A constitutive single crystal model as well as a homogenization scheme to account for the polycrystalline behavior of sheet metals is proposed. At the micro scale, crystal plasticity is used as constitutive model. In order to perform the micro-to-macro scale-bridging transition, a mean-field homogenization scheme will be used to account for the polycrystalline sheet-metal behavior. As preliminary studies have shown, Taylor type homogenization schemes do not suffice to predict the earing behavior accurately enough. Therefore, a non-linear Hashin-Shtrikman type homogenization scheme and its properties are investigated using forming simulations. Since the prediction of the resulting shape of structural parts in deep drawing simulations is the main objective, a numerically efficient implementation is crucial to keep computational times reasonable. In this context, the constitutive laws are implemented in the explicit FE-code which is especially well suited for the used material model and the homogenization scheme.

1.3 Outline

The necessary fundamentals of continuum mechanics, i.e., kinematics and balance equations, are briefly discussed in Section 2. Subsequently, the constitutive equations on the grain scale, for the geometrically linear as well as for the geometrically non-linear case, are discussed in Section 3. Accordingly, the numerical implementation of the material model is discussed in Section 4. The Hashin-Shtrikman homogenization scheme assuming constant stress polarizations is introduced in Section 5. Moreover, experimental results (performed by Constellium) in form of tensile tests, hydraulic bulge tests, texture measurements and earing profile measurements are presented in Section 6. The parameter identification of the material models with the assistance of the aforementioned experimental results is carried out in Section 7. Section 8 shows the results of the forming simulations and localization predictions. Finally, summary and conclusions are given in Section 9.

1.4 State of the art

The aim of this section is to give an overview on the main research topics linked to homogenization and two-scale simulations including a brief overview of the basic literature.

1.4.1 Homogenization schemes and data reduction procedures

Homogenization methods are utilized to predict the response of the heterogeneous material based on the material properties of the constituents. For linear material behavior, key works include the ones by Voigt (1889) and Reuss (1929), who assumed constant strains and stresses throughout the material, respectively. Eshelby (1957) solved the problem of a single inclusion in an infinite matrix and Hashin and Shtrikman (1962) obtained second-order bounds by using a variational principle. A more detailed overview on linear homogenization schemes is given in the works by Willis (1981), Mura (1987), Nemat-Nasser and Hori (1999), Torquato (2002), and Kanouté et al. (2009).

In the physically non-linear case, uniform strain is assumed by Taylor (1938) throughout a polycrystal (comparably to the Voigt approach in the aforementioned, linear case). This assumption provides an upper bound to the effective potential (Bishop and Hill, 1951) where the displacement field is kinematically compatible. In assuming overall uniform stresses kinematic compatibility is violated (see, e.g., Hosford and Galdos, 1990) while static compatibility is satisfied. This method of uniform stresses (lower bound) is often attributed to Sachs (Sachs, 1928), even though proportional and non-uniform loading was assumed by Sachs. In this context, proportional loading neither satisfies stress equilibrium nor strain compatibility (Leffers, 1979; Kocks et al., 1998). However, the term "Sachs-type behavior" for homogeneous stresses is commonly used in the literature and will be referred to as such in this work, too.

The aforementioned assumptions of homogeneous strains and stresses describe upper and lower bounds of the crystalline material behavior which, in general, is not present in real-life applications. Therefore, numerous attempts have been made to improve the homogenization resulting in estimates rather than bounds. Relaxed constrained theories (see, e.g., Kocks and Chandra, 1982; Van Houtte, 1982), also called relaxed Taylor theories, do not prescribe certain shear components of the strain tensor and thereby do not restrict the material behavior as much as in the Taylor model. Other approaches, like LAMEL (Van Houtte et al., 1999), ALAMEL (Van Houtte et al., 2005) or GIA models (Crumbach et al., 2001), connect neighboring domains (grains) which are subjected to macroscopic deformation. Similarly, the Sachs model, assuming homogeneous stresses and inducing only single-slip, has been modified by introducing random stresses in each grain and thereby inducing multi-slip (see, e.g., Leffers, 1979; Pedersen and Leffers, 1987).

An extension of the linear Hashin-Shtrikman scheme by assuming piecewise constant stress polarizations in every phase (or grain) and thereby making a non-linear stress response possible in every grain has been proposed by Jöchen and Böhlke (2012) and Jöchen (2013). A special case of this non-linear Hashin-Shtrikman scheme for elastically isotropic materials is formulated in Section 5.2 and applied to texture evolution, deep drawing and Nakajima simulations in Section 8.

In order to perform micro-mechanical FE simulations, the microstructure must be taken into account. For sheets the crystallographic texture is measured (by Constellium) and the resulting experimentally obtained orientation data or orientation distribution is too large for being handled numerically in a feasible way. Therefore, the orientation data set is reduced with the objective to still keep the texture representative. For this purpose, numerous methods are proposed in the literature.

In the work of Tóth and Van Houtte (1992), the orientation space is partitioned with a regular grid in order to create discrete orientations with identical volume fractions. Extending this method in terms of joining similar discrete orientations and weighting the volume fraction suitably, Melchior and Delannay (2006) increased the accuracy of texture approximations in their work.

Another method makes use of a characteristic set of cubic texture components such as cube, copper and brass components, which are represented by a von Mises-Fisher distribution with a specified half-width and volume

fraction being determined by merging orientations corresponding to a certain misorientation angle (Cho et al., 2004). Similarly, Raabe and Roters (2004) use characteristic metal texture components which are rotated in order to approximate the texture. The effect of too sharp texture representations compared to the uncompressed original, is counteracted there with random background texture components.
A mixed integer quadratic programming procedure is utilized in the work of Böhlke et al. (2006) to extract the main texture components with corresponding volume portions. In an optimization procedure, a small number of components is determined to represent the texture. This method also incorporates the introduction of background texture to smoothen overly sharp texture predictions.
Gao et al. (2006) used a partitioning of the orientation space with a suitable averaging procedure to obtain an orientation data set considerably reduced in size compared to the experimentally characterized original. An improvement of this method was suggested by Jöchen and Böhlke (2013), where a better suited fundamental zone choice in the orientation space is used, compared to the aforementioned method. This leads to improved numerics in the calculation of the data reduction and overall a more representative texture approximation of the original texture. Note that for the data reduction in Section 3.5 and throughout this work, this method is used exclusively.

1.4.2 Two-scale simulations and deep drawing

There is a strong demand to take the microstructure into account in FE simulations of structural parts, especially in metal forming operations, where several approaches have been persued (see, e.g., Geers et al., 2010). In this context, the approach with the highest accuracy is the so called FE^2 method (e.g., Renard and Marmonier, 1987; Miehe et al., 1999; Feyel, 2003). The microstructure, in this approach, is discretized with a RVE using finite elements and is coupled with the macroscopic

FE-discretization at the integration points. Therein, the macroscopic FE-model provides the local boundary conditions for the RVE FE-model on the micro-scale. So far, computational times for this method are far too long for the approach being used in cases other than academic examples, although the accuracy is tremendous (Van Houtte et al., 2012). One application, for instance, is the investigation of crystallographic texture on sheet metal failure in a limit dome height test (Nakamachi et al., 2007; Kuramae et al., 2010). Furthermore, as mentioned beforehand this method is very time consuming, that is why approaches have made to increase feasibility. A statistically similar RVE was introduced by Balzani et al. (2010) to accelerate the calculation at the microstructural level, whereas Novák et al. (2012) used a coarser mesh for the RVE, enabled by micromechanics-enhanced FE formulation.
Contrary to the FE^2 method, where the microstructure is represented by a RVE FE-problem, so called mean-field methods offer a computationally less expensive approach. They incorporate homogenization schemes rather then spatially discretizing the microstructure. In the Taylor method, at the Gauss point level, homogeneous strains in all crystals is assumed. Therewith, the deformation-induced crystallographic texture evolution can be modeled for rolling of sheet metals (Mathur et al., 1990; Aretz et al., 2000) and for steady state bulk forming (Mathur and Dawson, 1989; Kalidindi et al., 1992). Using the Taylor method in combination with an implicit FE code, Phan Van et al. (2012) modeled cup-forming of ferritic steel using deep drawing simulations. Combining FEs with the self-consistent (SC) scheme, as Tomé et al. (2001) and Segurado et al. (2012) did in an explicit and implicit FE-framework for the bending of a zirconium beam, leads to higher accuracy predictions. Additionally, the SC approach has been successfully applied to deep drawing of magnesium sheets (Walde and Riedel, 2007) and rolling of a FCC plate (Segurado et al., 2012). In order to even improve the computational efficiency, hierarchical multi-scale modeling in terms of identifying an anisotropic, phenomenological constitutive law by virtual experiments (Roters et al., 2010; Van Houtte

et al., 2012; Gawad et al., 2013; 2015) has been examined. In the hierarchical model, the texture evolution in terms of the plastic potential is updated locally if the deformation state requires a recomputation of the yield surface. The update is carried out on the microscale with virtual experiments based on crystal plasticity simulations (performed with Taylor and relaxed constrained Taylor models). Jöchen and Böhlke (2012) and Jöchen (2013) used a non-linear extension of the Hashin-Shtrikman scheme in an implicit FE code to perform deep drawing simulations of aluminum sheets. The aforementioned method is extended for the special case of elastic isotropy and plastic incompressibility in this work. This ansatz leads to a scalar homogenization parameter which, combined with an implementation in an explicit FE code, improves the numerical efficiency of the two-scale simulation. The method is applied to deep drawing simulations in Section 8.1.

Plastic anisotropy induced by crystallographic texture is a phenomenon frequently observed in engineering materials. The non-uniform cup height (earing) after deep drawing is a quality criterion in sheet metals, for example, as it may cause difficulties in beverage cans manufacured of aluminum (Blade, 1967; Hutchinson et al., 1989; Naess, 1991; Westerman, 1993; Ren and Das, 1998).

Forming operations have been simulated with FEM for a long time, where a phenomenological yield function has been used to model the plastic anisotropy of sheet metals (Gotoh and Ishise, 1978; Hill, 1990; Chung and Shah, 1992; Inal et al., 2000; Yoon et al., 2000). Although the implementation of these constitutive laws is often handy, the accuracy of these models is not always satisfying. For instance, many of these models cannot predict the occurance of six or eight ears, an issue that Yoon et al. (2006) solved by using the 18-parameter yield function of Barlat et al. (2005). Recently, Aretz and Barlat (2013) applied a phenomenological yield function for orthotropic sheet metals to the EN AW-3104 in deep drawing simulations. Although the prediction of the earing behavior is in good agreement with the experiments, a very large number (27) of parameters is needed to calibrate the material model. Unfortunately, the parameters of the yield function cannot be directly correlated to physical quantities. Also, the parameters cannot be calibrated by experiments only, as an additional optimization procedure is necessary to determine all of the 27 parameters. Another approach uses analytical yield functions, so called strain-rate potentials, instead of phenomenological ones. Therein, the yield locus can be determined with polycrystal-plasticity computations (e.g., Bacroix and Gilormini, 1995; Chung et al., 1996; Zhou et al., 1998; Li et al., 2001).

Earing behavior has been investigated by Engler and Hirsch (2007); Engler and Aegerter (2014), who applied the visco-plastic self-consistent scheme (Lebensohn and Tomé, 1993) and, thereby, incorporated the microstructure into their deep drawing simulations.

Utilizing crystal plasticity for earing predictions has been performed over the years for, e.g., deep drawing simulations and aluminum single crystal simulations (Becker et al., 1993), hydroforming of aluminum (Beaudoin et al., 1994), deep drawing of copper cups (Grujicic and Batchu, 2002) and the use of texture components for texture approximations in deep drawing of aluminum (Raabe and Roters, 2004). Crystal plasticity models have also been successfully applied to predict springback of metal sheets (geometric change of the part after release of forming forces). Schulze et al. (2009) incorporated texture information with a crystal plasticity material model for BCC sheet metals in deep drawing simulations with subsequent springback evaluation. In addition to predicting the earing behavior, some effort has been made to minimize the occurence of earing in deep drawn cups. For instance, Engler et al. (2011) used a texture-based polycrystal-plasticity model to predict earing and devise an improved blank shape to minimize it. A broad overview on crystal plasticity models and their various applications in material science is given in the review article of Roters et al. (2010).

1.5 Notation

A direct tensor notation is used throughout the text. Scalars and tensors of first-, second-, and fourth-order are denoted as a, $\boldsymbol{a}$, $\boldsymbol{A}$ and $\mathbb{A}$, respectively. The scalar and dyadic product of two tensors of arbitrary order is written as, e.g., $\boldsymbol{A} \cdot \boldsymbol{B}$ and $\boldsymbol{A} \otimes \boldsymbol{B}$, respectively. In addition, the Rayleigh product is defined as: $\boldsymbol{Q} \star \mathbb{A} = A_{ijkl}(\boldsymbol{Q}\boldsymbol{e}_i) \otimes (\boldsymbol{Q}\boldsymbol{e}_j) \otimes (\boldsymbol{Q}\boldsymbol{e}_k) \otimes (\boldsymbol{Q}\boldsymbol{e}_l)$. The composition of two second-order tensors is denoted as $\boldsymbol{AB}$. Moreover, a linear mapping of a second-order tensor by a forth-order tensor is denoted as $\boldsymbol{B} = \mathbb{C}[\boldsymbol{A}]$. Furthermore, $(\boldsymbol{A} \square \boldsymbol{B})\,[\boldsymbol{C}] = \boldsymbol{ABC}\ \forall \boldsymbol{A},\ \boldsymbol{B},\ \boldsymbol{C}$ is defined as well as the special contraction $(\boldsymbol{a} \otimes \boldsymbol{b}) \cdot (\mathbb{C}[[\boldsymbol{a} \otimes \boldsymbol{b}]]) = (\boldsymbol{a} \otimes \boldsymbol{a}) \cdot (\mathbb{C}[\boldsymbol{b} \otimes \boldsymbol{b}])$. The spherical and deviatoric part of a second-order tensor $\boldsymbol{A}$ are defined by $\boldsymbol{A}^\circ = \mathrm{tr}(\boldsymbol{A})/3$ and $\boldsymbol{A}' = \boldsymbol{A} - \boldsymbol{A}^\circ$, respectively.

1.6 Frequently used symbols and abbreviations

$(.)^n$	Quantity at time ^{n}
$(.)^{n+1}$	Quantity at time $^{n+1}$
$(.)^{\mathsf{T_L}}$	Transpose of left index pair (left minor transposition) for fourth-order tensors
$(.)^{\mathsf{T_R}}$	Transpose of right index pair (right minor transposition) for fourth-order tensors
$(.)^{\mathsf{T}}$	Transpose (major transposition for higher-order tensors) of Tensor quantity
$(\cdot)_0$	Initial/reference value of quantity
$(\cdot)_\alpha$	Slip system index, for FCC materials $\alpha = 1, \ldots, 12$
$(\cdot)_\infty$	Asymptotic value of quantity
$(\cdot)_\mathrm{e}$	Elastic quantity
$(\cdot)_\mathrm{p}$	Plastic quantity

$\Delta(.)$	Incremental quantity
$\bar{(.)}$	Effective (macroscopic) quantity
$\dot{(.)}$	Rate of quantity
$\langle . \rangle$	Volume average / Macaulay brackets
$\det(.)$	Determinant
$\mathrm{div}\,(.)$	Divergence
$\mathrm{Grad}\,(.)$	Lagrangian gradient
$\mathrm{grad}\,(.)$	Eulerian gradient
$\mathrm{sym}(.)$	Symmetric part
$\mathrm{skw}(.)$	Skew-symmetric part

Greek letters

γ	Accumulated plastic slip
Θ	Hardening modulus
λ^{HS}	Homogenization parameter $\lambda^{\mathrm{HS}} \in [0, 1]$
$\lambda,\ \mu$	Lamé constants
ν	Poisson's ratio
$\varphi_1,\ \Phi,\ \varphi_2$	Euler angles of crystallographic orientation in Bunge convention
ϱ	Density
τ_α	Schmid stress
τ^C	Critical resolved shear stress
τ^D	Drag stress
$\boldsymbol{\varepsilon}$	Strain tensor (infinitesimal)
$\boldsymbol{\tau}$	Kichhoff stress tensor
$\boldsymbol{\sigma}$	Cauchy stress tensor
$\boldsymbol{\omega}$	Rotation tensor (infinitesimal)

Latin letters

c_d	Dilatational wave speed
l_e	Characteristic element length
m	Hardening exponent
r	Lankford coefficient
t	Time
B_i	Box with index i for orientation reduction in Euler space
E	Young's modulus
G	Shear modulus
J	Jacobian
K	Bulk modulus
N	Number of boxes (data reduction method)
N_α	Number of slip systems
$SO(3)$	Group of all rotations about the origin in the three-dimensional space
V	Volume
$\boldsymbol{b}$	Body force
$\boldsymbol{d}$	Slip direction
$\boldsymbol{n}$	Normal direction
$\boldsymbol{t}$	Traction
$\boldsymbol{u}$	Displacement
$\boldsymbol{v}$	Velocity
$\boldsymbol{x}$ and $\boldsymbol{X}$	Position vector in current and in reference placement
$\boldsymbol{C}$	Right Cauchy Green Tensor
$\boldsymbol{D}$	Rate of deformation $(\mathrm{sym}(\boldsymbol{L}))$
$\boldsymbol{E}$	Green's strain tensor

$\boldsymbol{F}$	Deformation gradient
$\boldsymbol{H}$	Displacement gradient
$\boldsymbol{I}$	Identity on second-order tensors
$\boldsymbol{L}$	Velocity gradient
$\boldsymbol{M}_\alpha$	Schmid tensor of α-th slip system
$\boldsymbol{Q}$	Orientation tensor
$\boldsymbol{R}$	Rotation tensor
$\boldsymbol{S}$	Second Piola-Kirchhoff tensor
$\boldsymbol{U}$	Right stretch tensor
$\boldsymbol{W}$	Spin tensor ($\mathrm{skw}(\boldsymbol{L})$)
$\mathbb{A}$	Fourth-order strain localization tensor
$\mathbb{C}$	Fourth-order stiffness tensor
$\mathbb{C}_0$	Stiffness of comparison medium
$\mathbb{C}^{\mathrm{alg}}$	Algorithmic tangent operator
$\mathbb{I}$	Identity on fourth-order tensors
$\mathbb{P}_0$	Hill's polarization tensor
$\mathbb{P}_1^{\mathrm{iso}}$, $\mathbb{P}_2^{\mathrm{iso}}$	Isotropic fourth-order projectors

Abbreviations

Al	Aluminum
BCC	Body-centered cubic
Bs	Brass
CODF	Crystallite orientation distribution function
CPU	Central processing unit
Cu	Copper
EBSD	Electron backscatter diffraction
FCC	Face-centered cubic
FE	Finite element

FEM	Finite element method
FLC	Forming limit curve
FLD	Forming limit diagram
FZ	Fundamental zone
HS	Hashin-Shtrikman
ND	Normal direction (sheet metal)
PLC	Portevin-Le Chatelier
PSC	Plane strain compression
RC	Rotated cube orientation (22° around RD)
RD	Rolling direction (sheet metal)
RVE	Representative volume element
TD	Transverse direction (sheet metal)

Chapter 2

Kinematics and relevant balance equations

The fundamentals in this section are in line with the derivations from the textbooks Betten (2001), Haupt (2002) and Bertram (2008).

2.1 Kinematics

The position of a material point is described by the position vector $\boldsymbol{X}$ in the reference placement and the position vector $\boldsymbol{x} = \boldsymbol{\chi}(\boldsymbol{X}, t)$ in the current placement, available for all material points in a three-dimensional body at all times in the considered time interval. Moreover, the displacement field is introduced as the difference between the position vector in the current and the reference placement, i.e., $\boldsymbol{u}(\boldsymbol{X}, t) = \boldsymbol{\chi}(\boldsymbol{X}, t) - \boldsymbol{X}$. The material time derivative is defined by $\dot{\psi} = \partial \psi_L(\boldsymbol{X}, t)/\partial t$. Furthermore, the deformation gradient of a material point is introduced as the first spatial derivative of $\boldsymbol{\chi}$ with respect to $\boldsymbol{X}$

$$\boldsymbol{F} = \mathrm{Grad}(\boldsymbol{x}) = \frac{\partial \boldsymbol{\chi}(\boldsymbol{X}, t)}{\partial \boldsymbol{X}} \,. \tag{2.1}$$

Analogously, the displacement gradient is defined as

$$\boldsymbol{H} = \mathrm{Grad}(\boldsymbol{u}) = \frac{\partial \boldsymbol{u}(\boldsymbol{X}, t)}{\partial \boldsymbol{X}} = \boldsymbol{F} - \boldsymbol{I} \,, \tag{2.2}$$

with $\boldsymbol{I}$ being the identity on second-order tensors. The deformation gradient $\boldsymbol{F}$ accounts for both stretch and for rigid body rotations. Introducing the polar decomposition applicable to all invertible tensors

$$\boldsymbol{F} = \boldsymbol{R}\boldsymbol{U} \ , \tag{2.3}$$

the deformation gradient can be split into a symmetric positive-definite stretch tensor $\boldsymbol{U}$ and a proper-orthogonal rotation tensor $\boldsymbol{R}$. Furthermore, the velocity gradient can be defined as

$$\boldsymbol{L} = \mathrm{grad}(\boldsymbol{v}) = \frac{\partial \boldsymbol{v}(\boldsymbol{x}, t)}{\partial \boldsymbol{x}} = \dot{\boldsymbol{F}}\boldsymbol{F}^{-1} \ , \tag{2.4}$$

with grad being an Eulerian gradient. The velocity gradient can be additively decomposed into a symmetric part $\boldsymbol{D}$ (rate of deformation) and a skew-symmetric part $\boldsymbol{W}$ (spin tensor)

$$\boldsymbol{L} = \mathrm{sym}(\boldsymbol{L}) + \mathrm{skw}(\boldsymbol{L}) = \boldsymbol{D} + \boldsymbol{W} \ . \tag{2.5}$$

Strain measures are used to describe the change of line elements in terms of length and angle. Several strain tensors can be defined, e.g., Biot strain, Hencky stain, etc. (Bertram, 2008). Another common strain measure is Green's strain tensor

$$\boldsymbol{E} = \frac{1}{2}\left(\boldsymbol{F}^{\mathsf{T}}\boldsymbol{F} - \boldsymbol{I}\right) = \frac{1}{2}\left(\boldsymbol{H} + \boldsymbol{H}^{\mathsf{T}} + \boldsymbol{H}^{\mathsf{T}}\boldsymbol{H}\right) \ , \tag{2.6}$$

which, similar to all generalized strain measures, takes the form of the infinitesimal strain tensor $\boldsymbol{\varepsilon}$ for small deformations. The infinitesimal strain $\boldsymbol{\varepsilon}$ and the rotation tensor $\boldsymbol{\omega}$ can be denoted in terms of the symmetric and the skew-symmetric part of the displacement gradient, respectively:

$$\boldsymbol{\varepsilon} = \frac{1}{2}\left(\boldsymbol{H} + \boldsymbol{H}^{\mathsf{T}}\right) = \mathrm{sym}(\boldsymbol{H}) \quad , \quad \boldsymbol{\omega} = \frac{1}{2}\left(\boldsymbol{H} - \boldsymbol{H}^{\mathsf{T}}\right) = \mathrm{skw}(\boldsymbol{H}) \ . \tag{2.7}$$

The infinitesimal strain tensor is valid if a geometrical linearization can be performed, i.e., if $\| \boldsymbol{H} \| \ll 1$ holds.

2.2 Relevant balance equations

The balance of linear momentum states that the time derivative of linear momentum is equal to the external forces. With the mass density ϱ, the body forces $\boldsymbol{b}$, and the Cauchy stress tensor $\boldsymbol{\sigma}$, the local form of the balance equation in regular points reads as

$$\varrho \ddot{\boldsymbol{u}} = \operatorname{div}(\boldsymbol{\sigma}) + \varrho \boldsymbol{b} \,. \tag{2.8}$$

The linear momentum is formulated for the solution with the FE method in a weak form and reads (Wriggers, 2008)

$$\int_{\mathcal{B}} \boldsymbol{\sigma} \cdot \operatorname{Grad}(\boldsymbol{w}) \, \mathrm{d}V - \int_{\mathcal{B}} \varrho(\boldsymbol{b} - \ddot{\boldsymbol{u}}) \cdot \boldsymbol{w} \, \mathrm{d}V - \int_{\partial\mathcal{B}^{\sigma}} \boldsymbol{t} \cdot \boldsymbol{w} \, \mathrm{d}A = 0 \tag{2.9}$$

with Cauchy's theorem $\boldsymbol{t} = \boldsymbol{\sigma n}$ and the test function $\boldsymbol{w}$.
Due to the balance of angular momentum for non-polar media, the symmetry of the stress tensor holds

$$\boldsymbol{\sigma} = \boldsymbol{\sigma}^{\mathsf{T}} \,. \tag{2.10}$$

In case of static equilibrium, the left-hand side of eq. (2.8) is equal to zero. The boundary $\partial\mathcal{B}$ of a body $\mathcal{B}$ can be divided into a Dirichlet boundary $\partial\mathcal{B}^{u}$ (displacement $\boldsymbol{u} = \bar{\boldsymbol{u}}$ is prescribed) and a Neumann boundary $\partial\mathcal{B}^{\sigma}$ (traction $\boldsymbol{t} = \boldsymbol{\sigma n} = \bar{\boldsymbol{t}}$ is prescribed) with

$$\partial\mathcal{B}^{\sigma} \cap \partial\mathcal{B}^{u} = \emptyset \,, \qquad \partial\mathcal{B}^{\sigma} \cup \partial\mathcal{B}^{u} = \partial\mathcal{B} \,. \tag{2.11}$$

Chapter 3

Constitutive equations on the grain scale

3.1 Geometrically linear formulation

The infinitesimal strain tensor ε can be split into a reversible, elastic part ε_e and an inelastic part ε_p

$$\varepsilon = \varepsilon_\mathrm{e} + \varepsilon_\mathrm{p} \ . \tag{3.1}$$

The Cauchy stress tensor $\boldsymbol{\sigma}$ is given by a linear mapping of the strain tensor ε_e by the stiffness tensor $\mathbb{C}$ (Hooke's law): $\boldsymbol{\sigma} = \mathbb{C}[\varepsilon_\mathrm{e}]$. In rate form, Hooke's law reads

$$\dot{\boldsymbol{\sigma}} = \mathbb{C}[\dot{\varepsilon}_\mathrm{e}] \ . \tag{3.2}$$

Due to the elastically isotropic behavior of aluminum, the isotropic fourth-order stiffness tensor $\mathbb{C}$ can be used: $\mathbb{C} = \lambda \boldsymbol{I} \otimes \boldsymbol{I} + 2\mu \mathbb{I}^S$, with the Lamé constants λ and μ, and the symmetric part of fourth-order identity $\mathbb{I}^S$. The evolution equation for the inelastic part of the strain tensor is given by the flow rule (see e.g., Asaro, 1983; Bertram, 2008)

$$\dot{\varepsilon}_\mathrm{p} = \sum_{\alpha=1}^{N_\alpha} \dot{\gamma}_\alpha \, \mathrm{sym}(\boldsymbol{M}_\alpha) \ , \tag{3.3}$$

with the Schmid tensor $\boldsymbol{M}_\alpha = \boldsymbol{d}_\alpha \otimes \boldsymbol{n}_\alpha$, slip plane direction $\boldsymbol{d}$ and the slip plane normal $\boldsymbol{n}$ of each slip system α. In case of face-centered cubic (FCC) crystal symmetry $N_\alpha = 12$ (four slip planes with three slip directions each). Furthermore, the evolution of plastic slip in the α-th slip system can be given by an overstress relation (see e.g., Méric et al., 1994)

$$\dot{\gamma}_\alpha = \dot{\gamma}_0 \mathrm{sgn}\,(\tau_\alpha) \left\langle \frac{|\tau_\alpha| - \tau^C}{\tau^D} \right\rangle^m , \tag{3.4}$$

with the Schmid stress $\tau_\alpha = \boldsymbol{\sigma} \cdot \mathrm{sym}(\boldsymbol{M}_\alpha)$, the critical resolved shear stress τ^C, the drag stress τ^D, and the reference slip rate $\dot{\gamma}_0$. The Macaulay brackets in eq. (3.4) are defined as $\langle x \rangle = (|x| + x)/2$. Isotropic hardening is often modeled with a Voce-type exponential saturation function (Voce, 1955). Here, this approach is combined with an additional linear term $\Theta_\infty \gamma$ (e.g., Jain et al., 1996)

$$\tau^C = \tau_0^C + \Theta_\infty \gamma + \left(\tau_\infty^C - \tau_0^C\right) \left(1 - \exp\left(-\frac{\Theta_0 - \Theta_\infty}{\tau_\infty^C - \tau_0^C}\gamma\right)\right) , \tag{3.5}$$

with the initial and the asymptotic values for the critical resolved shear stress (τ_0^C , τ_∞^C), the hardening moduli (Θ_0 , Θ_∞), and the accumulated plastic slip $\dot{\gamma} = \sum_\alpha^{N_\alpha} |\dot{\gamma}_\alpha|$.

3.2 Incremental form of the geometrically linear formulation

An incremental form of eq. (3.1) is given by

$$\Delta\varepsilon = \Delta\varepsilon_\mathrm{e} + \Delta\varepsilon_\mathrm{p} \, . \tag{3.6}$$

Hence, the incremental form of the Hooke's law reads

$$\Delta\boldsymbol{\sigma} = \mathbb{C}[\Delta\varepsilon - \Delta\varepsilon_\mathrm{p}] \, , \tag{3.7}$$

where the inelastic part of the incremental strain tensor is given by the following relation

$$\Delta\varepsilon_{\mathrm{p}} = \sum_{\alpha=1}^{N_\alpha} \Delta\gamma_\alpha \,\mathrm{sym}(\boldsymbol{M}_\alpha)\,. \tag{3.8}$$

In a given time increment $\Delta t = t^{n+1} - t^n$, the deformation gradients at the beginning, $\boldsymbol{F}^n$, and at the end, $\boldsymbol{F}^{n+1}$, are assumed to be given. Then, the deformation during this time increment is $\Delta\boldsymbol{F} = \boldsymbol{F}^{n+1}(\boldsymbol{F}^n)^{-1}$. In metal forming operations, a widely used measure of deformation is the Hencky strain $\boldsymbol{E}^H = 1/2\ln(\boldsymbol{C}) = \ln(\boldsymbol{U})$. Expanding the Hencky strain into a Taylor series and considering the incremental form, a second-order approximation of the small strain measure is obtained:

$$\Delta\varepsilon \cong \frac{1}{2}\left(\Delta\boldsymbol{F}^{\mathsf{T}}\Delta\boldsymbol{F} - \boldsymbol{I}\right) + \frac{1}{4}\left(\Delta\boldsymbol{F}^{\mathsf{T}}\Delta\boldsymbol{F} - \boldsymbol{I}\right)^2\,, \tag{3.9}$$

which is incrementally objective, i.e., pure rotations do not induce any pseudo strains. Furthermore, with eq. (3.4), the slip increment of the α-th slip system is

$$\Delta\gamma_\alpha = \Delta t\dot{\gamma}_0 \mathrm{sgn}\left(\tau_\alpha\right)\left\langle\frac{|\tau_\alpha| - \tau^C}{\tau^D}\right\rangle^m\,. \tag{3.10}$$

The accumulated incremental slip is $\Delta\gamma = \sum_\alpha^{N_\alpha}|\Delta\gamma_\alpha|$. As for the strain increments, an additive split is valid also for the incremental rotations:

$$\Delta\boldsymbol{\omega} = \Delta\boldsymbol{\omega}_{\mathrm{e}} + \Delta\boldsymbol{\omega}_{\mathrm{p}}\,, \tag{3.11}$$

with the plastic rotation increment being obtained from the skew part of the Schmid tensor and the slip increment as

$$\Delta\boldsymbol{\omega}_{\mathrm{p}} = \sum_{\alpha=1}^{N_\alpha} \Delta\gamma_\alpha \mathrm{skw}(\boldsymbol{M}_\alpha)\,. \tag{3.12}$$

From eq. (2.4), one can obtain the differential equation $\dot{\boldsymbol{F}} = \boldsymbol{L}\boldsymbol{F}$ which leads to an objective approximation of the velocity gradient for an implicit Euler scheme. With the relation $\Delta\boldsymbol{\omega} = \Delta t\boldsymbol{W} = \Delta t\,\mathrm{skw}(\boldsymbol{L})$, the total spin increment reads (see e.g., Simo and Hughes, 1998; Miehe et al., 2010)

$$\Delta\boldsymbol{\omega} = \frac{1}{2}(\Delta\boldsymbol{F}^{-\mathsf{T}} - \Delta\boldsymbol{F}^{-1})\;, \tag{3.13}$$

with the skew-symmetric form of the equation ensuring that pure stretches do not induce rotations. Moreover, with the differential equation for $\dot{\boldsymbol{Q}}\boldsymbol{Q}^{-1} = \dot{\boldsymbol{\omega}}_{\mathrm{e}}$, the lattice rotation $\boldsymbol{Q}^{n+1}$ can be determined using the exponential mapping

$$\boldsymbol{Q}^{n+1} = \exp{(\Delta\boldsymbol{\omega}_{\mathrm{e}})}\boldsymbol{Q}^{n}\;, \tag{3.14}$$

with $\Delta\boldsymbol{\omega}_{\mathrm{e}} = \Delta\boldsymbol{\omega} - \Delta\boldsymbol{\omega}_{\mathrm{p}}$.
The incremental rotation is given by

$$\Delta\boldsymbol{Q} = \boldsymbol{Q}^{n+1}(\boldsymbol{Q}^{n})^{-1} = \exp{(\Delta\boldsymbol{\omega}_{\mathrm{e}})}\;. \tag{3.15}$$

With the computed lattice rotation, in every step of the computation, the Eulerian Schmid tensor $\boldsymbol{M}^{n}_{\alpha} = \boldsymbol{Q}^{n}\tilde{\boldsymbol{M}}_{\alpha}\boldsymbol{Q}^{n\mathsf{T}}$ is updated.

3.3 Geometrically non-linear formulation

A multiplicative decomposition of the deformation gradient is assumed as follows

$$\boldsymbol{F} = \boldsymbol{F}_{\mathrm{e}}\boldsymbol{F}_{\mathrm{p}}\;, \tag{3.16}$$

with the elastic and inelastic part, $\boldsymbol{F}_{\mathrm{e}}$ and $\boldsymbol{F}_{\mathrm{p}}$, respectively (Asaro and Needleman, 1985; Böhlke et al., 2005).
For the determinant of the inelastic part of the Jacobian $J_p = \det(\boldsymbol{F}_{\mathrm{p}}) = 1$ is assumed (plastic incompressibility), which implies that the determinant of the deformation gradient and is equal to the determinant of the elastic

deformation gradient: $J = \det(\boldsymbol{F}) = \det(\boldsymbol{F}_{\mathrm{e}}) = J_e$. The elastic behavior is modeled here with the St. Venant law

$$\boldsymbol{S}_e = \mathbb{C}[\boldsymbol{E}_e] \; . \tag{3.17}$$

Therein, the second Piola-Kirchhoff tensor $\boldsymbol{S}_e$ and elastic Green strain tensor $\boldsymbol{E}_e = (\boldsymbol{F}_{\mathrm{e}}{}^{\mathsf{T}}\boldsymbol{F}_{\mathrm{e}} - \boldsymbol{I})/2$ (see eq. (2.6)). Hence, the Kirchhoff stress is

$$\boldsymbol{\tau} = \boldsymbol{F}_{\mathrm{e}}\boldsymbol{S}_e\boldsymbol{F}_e^{\mathsf{T}} \; . \tag{3.18}$$

The evolution equation for the elastic part of the deformation gradient reads

$$\dot{\boldsymbol{F}}_e = \boldsymbol{L}\boldsymbol{F}_{\mathrm{e}} - \boldsymbol{F}_{\mathrm{e}}\boldsymbol{L}_p \; , \tag{3.19}$$

with the velocity gradient $\boldsymbol{L}$ according to eq. (2.4) and the inelastic part of the velocity gradient $\boldsymbol{L}_p = \dot{\boldsymbol{F}}_p\boldsymbol{F}_p^{-1}$. The following ansatz is used for $\boldsymbol{L}_p$:

$$\boldsymbol{L}_p = \sum_{\alpha}^{N_\alpha} \dot{\gamma}_\alpha \boldsymbol{M}_\alpha \; , \tag{3.20}$$

where the evolution of slip of the α-th slip system is $\dot{\gamma}_\alpha$ (see eq. (3.4)) and the Schmid tensor is $\boldsymbol{M}_\alpha$. The accumulated slip rate reads

$$\dot{\gamma} = \sum_{\alpha}^{N_\alpha} |\dot{\gamma}_\alpha| \; . \tag{3.21}$$

In the large deformation setting, the Schmid stress from eq . (3.4) is

$$\tau_\alpha = \boldsymbol{C}_e\boldsymbol{S}_e \cdot \boldsymbol{M}_\alpha \cong \boldsymbol{S}_e \cdot \boldsymbol{M}_\alpha \; , \tag{3.22}$$

with the elastic right Cauchy Green Tensor $\boldsymbol{C}_e = \boldsymbol{F}_{\mathrm{e}}{}^{\mathsf{T}}\boldsymbol{F}_{\mathrm{e}}$, and the elastic second Piola-Kirchhoff stress tensor $\boldsymbol{S}_e$. For small elastic deformations, the approximation $\boldsymbol{C}_e \cong \boldsymbol{I}$ can be used in eq. (3.22). Analogously to the linear framework, the hardening is modeled with a Voce-type law, see eq. (3.5). The initial conditions at $t = 0$ are as follows:

$$\boldsymbol{F}_{\mathrm{e}}(0) = \boldsymbol{Q}_0 \in SO(3) \qquad \boldsymbol{Q}_0 = \boldsymbol{g}_i(0) \otimes \boldsymbol{e}_i \qquad \gamma(t=0) = 0 \,. \tag{3.23}$$

Therein, the vectors $\boldsymbol{g}_i(t)$ describe the local base vectors and, thereby, the orientation $\boldsymbol{Q}$ at time t.

3.4 Representation of crystallographic orientations

In order to describe the polycrystalline material behavior by a single-crystalline constitutive law, the orientation of each crystal needs to be considered appropriately. There are different mathematically equivalent ways to describe an orientation, e.g., quaternions or axis-angle notation. Here, the description of orientations with orthogonal tensors and Euler angles is used. In the Bunge convention (Bunge, 1965; Bunge and Esling, 1982) the crystal orientation is described by three rotations of an initial crystal orientation. First, a rotation around the reference z-axis is imposed, followed by a rotation around the new x-axis, and then finally by a rotation around the new z-axis. The angles describing these rotations are denoted by φ_1, Φ, φ_2, respectively, with the following domains:

$$0 \le \varphi_1 < 2\pi \,, \quad 0 \le \Phi < \pi \,, \quad 0 \le \varphi_2 < 2\pi \,. \tag{3.24}$$

The z, x, z-rotation with respect to the reference coordinate system can be expressed by the rotation tensor

$$\boldsymbol{Q}(\varphi_1, \Phi, \varphi_2) = \begin{pmatrix} c_{\varphi_1} c_{\varphi_2} - s_{\varphi_1} c_\Phi s_{\varphi_2} & -c_{\varphi_1} s_{\varphi_2} - s_{\varphi_1} c_\Phi c_{\varphi_2} & c_{\varphi_1} c_\Phi \\ s_{\varphi_1} c_{\varphi_2} - c_{\varphi_1} c_\Phi s_{\varphi_2} & -s_{\varphi_1} s_{\varphi_2} + c_{\varphi_1} c_\Phi c_{\varphi_2} & -c_{\varphi_1} c_\Phi \\ s_\Phi s_{\varphi_2} & s_\Phi c_{\varphi_2} & c_\Phi \end{pmatrix} , \tag{3.25}$$

with the components of the rotation tensor abbreviated with the introduction of $c_\alpha := \cos(\alpha)$ and $s_\alpha := \sin(\alpha)$.

The symmetrically equivalent orientations due to crystal and sample symmetry are defined through

$$\boldsymbol{Q}^{eq} = \boldsymbol{Q}^S \boldsymbol{Q} \boldsymbol{Q}^C \,, \quad \forall\, \boldsymbol{Q}^C \in S^C \subseteq SO(3) \,, \quad \forall\, \boldsymbol{Q}^S \in S^S \subseteq SO(3) \,. \tag{3.26}$$

Here, S^C and S^S represent the symmetry group of the crystal (e.g., cubic crystal symmetry for aluminum) and the sample (e.g., orthotropic symmetry for ideal rolling textures with orthorhombic symmetry).

The crystallite orientation distribution function (CODF) has been proposed to characterize the orientation distribution in a polycrystal. Bunge (1965) defines $f(\boldsymbol{Q})$ as the CODF with $\mathrm{d}v/v$, the volume fraction of one crystal in the polycrystal, $\boldsymbol{Q} + \mathrm{d}\boldsymbol{Q}$, the appropriate orientation and $\mathrm{d}\boldsymbol{Q}$ the volume element in $SO(3)$ as

$$\frac{\mathrm{d}v}{v}(\boldsymbol{Q}) = f(\boldsymbol{Q})\,\mathrm{d}\boldsymbol{Q} \,. \tag{3.27}$$

Furthermore, the CODF accounts for the sample symmetry as well as the crystal symmetry of the polycrystalline aggregate and is both non-negative and normalized (Bunge, 1982)

$$\int\limits_{SO(3)} f(\boldsymbol{Q})\,\mathrm{d}\boldsymbol{Q} = 1 \,. \tag{3.28}$$

For a more detailed look on this topic the interested reader is referred to Bunge (1982); Engler and Randle (2010).

3.4.1 Pole figures and contour plots of CODF sections

The texture data can be visualized with, e.g., pole figures or CODF sections. The fundamentals of the methods presented here are in line with the derivations in Bunge (1982); Engler and Randle (2010); Schwartz et al.

(2009). Pole figures are a widespread method to visualize orientation data and textures. In Fig. 3.1, the stereographical projection is shown, where the piercing point of the normals of the crystallographic planes with the upper hemisphere of a unit spere is projected (through the south pole of the sphere as reference point) onto the equatorial plane of the unit sphere. The (100), (110) and (111) pole figures of one crystal have different numbers of poles (three, six and four, respectively) resulting from the number of crystallographic planes and the associated normals (see Fig. 3.2).

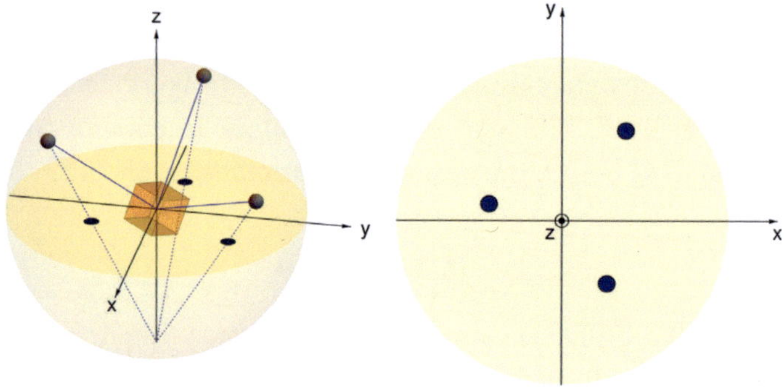

Figure 3.1: Stereographic projection (left) and (100) pole figure of the orientation with Euler angles $\varphi_1 = 35°$, $\Phi = 50°$ and $\varphi_2 = 30°$

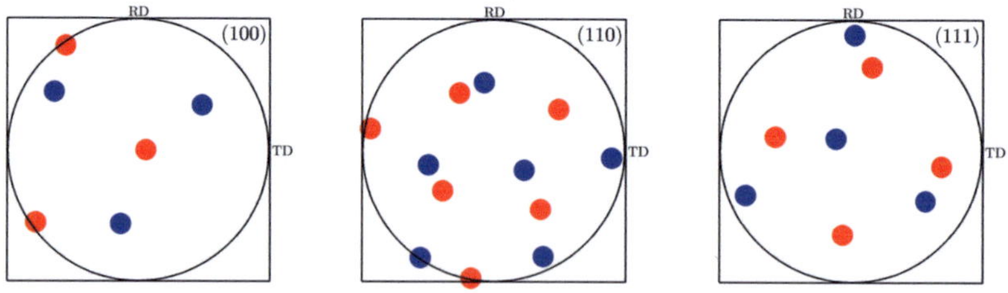

Figure 3.2: Pole figures (100), (110) and (111) of a cubic single crystal: Two orientations with Euler angles $\varphi_1 = 0°$, $\Phi = 5°$, $\varphi_2 = 35°$ (●) and $\varphi_1 = 35°$, $\Phi = 50°$, $\varphi_2 = 30°$ (●)

For materials with a pronounced texture, there is a concentration of poles in the pole figures, i.e., clusters of orientations are formed. This can be seen for a cold-rolled aluminum alloy (EN AW-3104 from Section 6.7) in form of a so-called rolling-induced texture in Fig. 3.3.

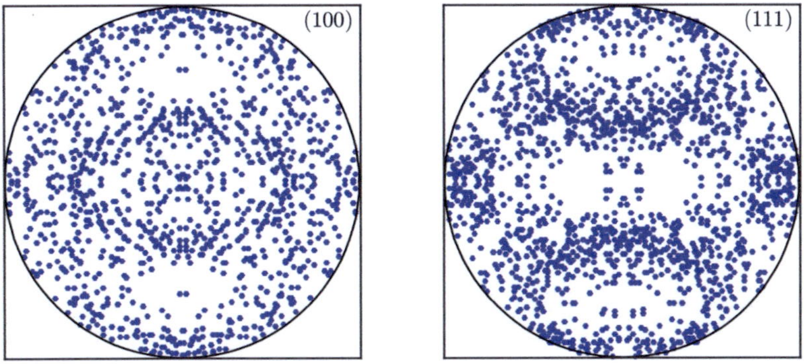

Figure 3.3: Discrete pole figures of aluminum EN AW-3104 (see Section 6.7): (100) (left) and (111) (right) direction

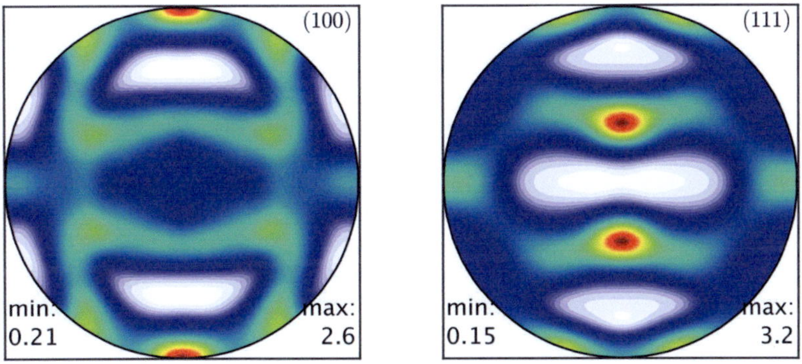

Figure 3.4: Continuous pole figures of EN AW-3104 (see Section 6.7): (100) (left) and (111) (right) direction

Pole figures with continuous density distribution of orientations can exhibit different volume fractions of crystals. This is contrary to discrete pole figures where crystallographically equivalent orientations in terms of their weights cannot be distinguished anymore. In order to generate a continuous pole figure plot, each of the aforementioned crystals in $SO(3)$ is described by a continuous function, a so called kernel function (e.g., de La Vallée Poussin (Schaeben, 1997)). The pole figures of a rolling texture from Fig. 3.3 (discrete) are shown as continuous pole figures in Fig. 3.4. Another method for displaying orientations is shown in Fig. 3.6, where a cartesian coordinate system with the three Euler angles φ_1, Φ and φ_2 as coordinates. The three-dimensional orientation space is depicted, here, in two-dimensionial sections with φ_2 = constant in Fig. 3.5. There, the typical texture components for a rolling texture of an aluminum alloy are shown with their location in the orientation space. The texture components of Cube, Goss, S, Brass and Copper are taken from Engler and Randle (2010) and are summarized in Tab. 3.1.

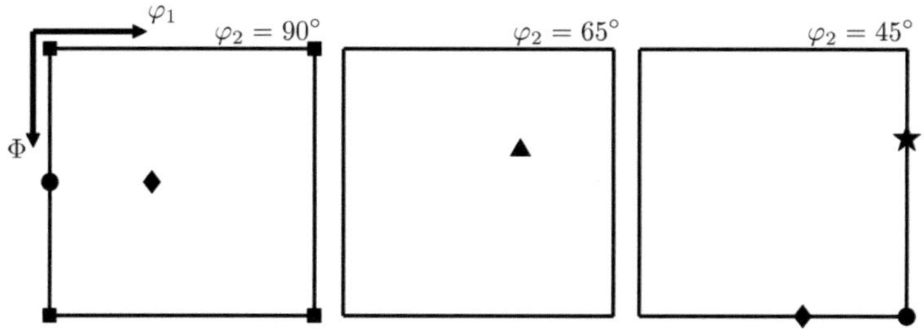

Figure 3.5: Characteristic texture components of aluminum Cube ■, Goss ●, Bs ♦, Cu ★ and S ▲ (see Tab. 3.1): Sections in Euler space with constant $\varphi_1 = 90°, 65°, 45°$

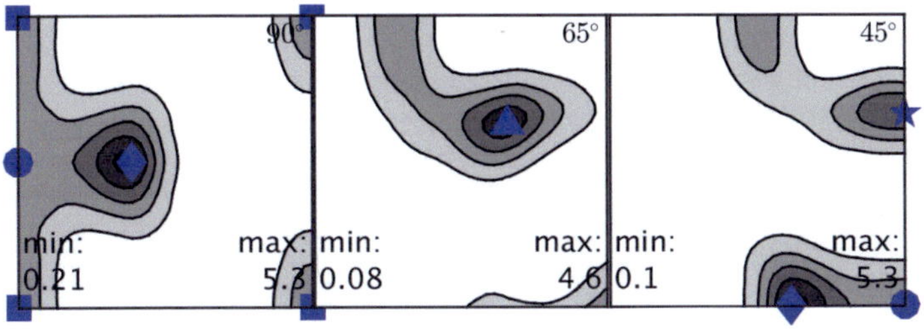

Figure 3.6: Orientation distribution by contour plot of CODF in φ_2-sections in Euler space of EN AW-3104 (see Section 6.7) with contours at 1, 2, 3, 4, 5 and 6

In Fig. 3.6, φ_2-sections of the Euler space are shown for the aluminum alloy (EN AW-3104) with the characteristic orientations from Tab. 3.1 indicated additionally. The characteristic texture components are in very good agreement with the cold rolled texture. Throughout this work, the software MTEX (Hielscher and Schaeben, 2008; Mainprice et al., 2011) is used for evaluating pole figures and CODF, the software is redistributed through Schaeben and Hielscher (2016). Furthermore, in the follwing the de La Vallée Poussin kernel (Schaeben, 1997) will be used for the CODF with a halfwidth of 6°.

Table 3.1: Euler angles of most important orientations of aluminum alloys after rolling (approximated) taken from Engler and Randle (2010); RC (22° rotated cube component) orientation taken from Delannay (2001)

Designation	Euler angles			Designation	Euler angles		
	φ_1	Φ	φ_2		φ_1	Φ	φ_2
Cube	0°	0°	0°	S	59°	34°	65°
Goss	0°	45°	0°	Brass (Bs)	35°	45°	0°
RC	22°	0°	0°	Copper (Cu)	90°	30°	45°

3.5 Data reduction method for crystallographic orientations

The measurement of crystallographic texture usually leads to a large amount of orientation information. With EBSD, a sector of the grain structure is characterized and experimental data can be gathered regarding microstructure, texture and grain morphology (Engler and Randle, 2010). In addition, with the X-ray diffraction method an orientation density function (ODF) can be measured.
Contrary to the EBSD method, however, morphology information is lost and the discrete set of orientations is not accessible. In order to use the micromechanical model at the Gauss-point level, the method is implemented in the commercial FE-code Abaqus. Due to the constrained number of solution-dependent state variables in the Abaqus user interface (10000 state variables is the limit as stated in Simulia (2012)) and in order to account for numerically feasible calculation times of parts, the orientation dataset needs to be reduced.
To perform such a reduction, the orientation data needs to be transformed into fundamental zones (FZ). The Euler space with the 24 cubic fundamental zones (eight sub-cuboids with three FZ each) is shown in Fig. 3.7 (left), where every sub-cuboid has the dimensions of $\varphi_1 = 2\pi$, $\Phi = \pi/2$ and $\varphi_2 = \pi/2$ (see, Jöchen and Böhlke, 2013). For orthorhombic sample symmetry, as in the case of rolled sheets considered subsequently, there are $24 \times 4 = 96$ FZ. For this case, the limit in φ_1-direction decreases to $\pi/2$ in comparison to the case of 24 fundamental zones. A data-reduction technique for crystallographic orientations has been introduced by Gao et al. (2006). The main texture components are obtained by transforming the orientation in FZ depending on the crystallographic symmetry class. The orientation space of the fundamental zone is then divided into subzones (called tessellations) and averaged taking account the metrics of the orientation space (see Fig. 3.8 (right)). The method was further developed

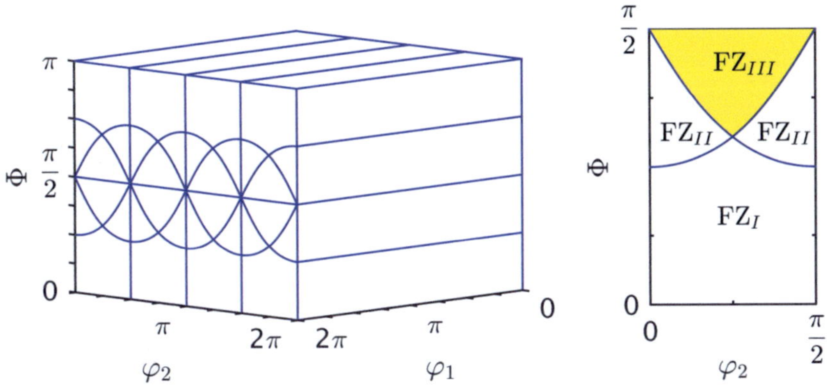

Figure 3.7: Euler space: Full cartesian Euler space with cubic fundamental zones (left) and cubic fundamental zones FZ_I, FZ_{II} and the chosen fundamental zone for the data reduction FZ_{III} highlighted in yellow (right)

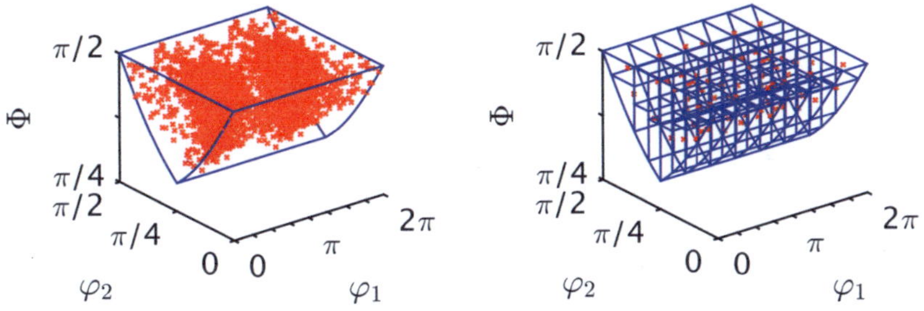

Figure 3.8: Orientations of EN AW-3104 (see chapter 6.7) in FZ: original data set (left) and reduced data set with $N_I = 6$, $N_J = 2$ and $N_K = 6$ (right)

by Jöchen and Böhlke (2013) in order to account for numerical aspects in the calculation of the tessellations by choosing another FZ (FZ_{III} in Fig. 3.7 (right)) than in the previous works. This method is summarized briefly in the following.

The boxes are being formed with the constraint of equal volume cells in the Euler orientation space and taking into account the sample symmetry

$$V(B_i) = \frac{1}{8\pi^2 N} \int\limits_{\text{box}} \sin\Phi \, \mathrm{d}\varphi_1 \, \mathrm{d}\Phi \, \mathrm{d}\varphi_2 = \frac{1}{F_{\text{sym}} N}, \qquad F_{\text{sym}} = \{24; 96\} \, . \tag{3.29}$$

For a fixed and prescribed number of boxes their boundaries are determined in an iterative process. In φ_1-direction,

$$0 \leq \varphi_1 < \begin{cases} 2\pi & \text{triclinic sample symmetry} \\ \pi/2 & \text{orthotropic sample symmetry} \end{cases} , \tag{3.30}$$

the orientation space can be subdivided equidistantly with the prescribed number of boxes in this direction. In the other directions, the nonlinear bounding of the FZ domain

$$\arccos\left(\frac{\cos(\varphi_2 + 3\pi/2)}{\sqrt{1+\cos^2(\varphi_2)}}\right) \leq \Phi \leq \frac{\pi}{2} \quad \forall \quad \varphi_2 \in \left[0, \frac{\pi}{4}\right] , \tag{3.31}$$

$$\arccos\left(\frac{\cos(\varphi_2)}{\sqrt{1+\cos^2(\varphi_2)}}\right) < \Phi < \frac{\pi}{2} \quad \forall \quad \varphi_2 \in \left(\frac{\pi}{4}, \frac{\pi}{2}\right) , \tag{3.32}$$

must be taken into account with $0 \leq \varphi_2 < \pi/2$. Recursive computations must be performed in order to establish the limits of the boxes (Jöchen and Böhlke, 2013). The crystallographic texture from Section 7.2 can be found in Tab. 7.2 (72×4) with N_I, N_J and N_K being the amount of boxes in φ_1-, Φ- and φ_2-direction (two different examples can be found in Fig. 3.9, wherein the right figure represents the tessellation in Fig. 3.8).

In Fig. 3.10, the volume fraction of the data set (EN AW-3104) in Fig. 3.8 (right) is shown. There, for instance, the boxes with the highest volume fractions (box ID 17, 19, 53, 55) correspond to the characteristic Brass component (see Fig. 3.5 and Tab. 3.1) and consequently produce the highest intensity level in the CODF (see Fig. 3.6).

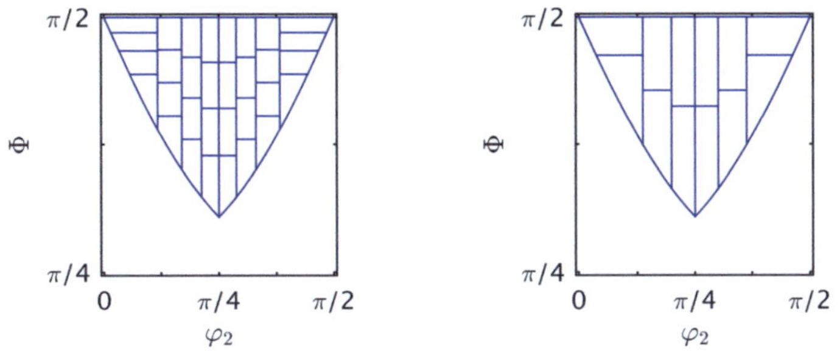

Figure 3.9: Partition of FZ_{III} with tessellation method: subdivisions with $N_J = 4$, $N_K = 8$ (left) and $N_J = 2$, $N_K = 6$ (right)

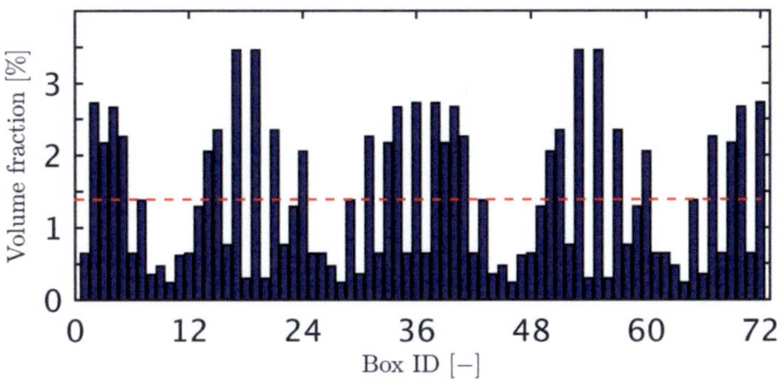

Figure 3.10: Statistics of orientation dataset with $N_I = 6$, $N_J = 2$ and $N_K = 6$: volume fraction of orientations per box with mean value

Chapter 4

Numerical implementation

4.1 Implementation of ABAQUS user routines

4.1.1 ABAQUS Explicit (VUMAT)

The VUMAT (Vectorized User MATerial) interface is used to numerically integrate the constitutive equations in ABAQUS Explicit. The Fortran subroutines are evaluated for blocks of data and are vectorized, which improves the numerical performance. The stresses and internal variables need to be updated with an appropriate integration scheme. The stresses in the VUMAT are calculated in the corotational frame. ABAQUS automatically accounts for rigid body motions and transforms the stresses into the current configuration. Contrary to ABAQUS Standard, the algorithmic tangent operator does not need to be computed (Simulia, 2012).

4.1.2 ABAQUS Standard (UMAT)

In ABAQUS Standard, an implicit finite element code is used for applications, for instance, in structural mechanics. A system of equations is solved incrementally. The ABAQUS UMAT (User MATerial) interface can be used to define the mechanical behavior of a material. The constitutive and evolution equations are integrated, which is performed with an implicit Euler scheme at every integration point of each finite element.

For a prescribed deformation, the stresses, internal variables, and algorithmic tangent operator (see, e.g., Simo and Hughes, 1998) need to be updated with the UMAT interface (Simulia, 2012). For the case of large-strain crystal plasticity, the integration algorithm of the internal variables is presented in Section 4.2.2.

4.2 Numerical time integration of the constitutive equations

4.2.1 Incremental formulation: explicit time integration

In an explicit finite element setting, the unknowns for the new time step t^{n+1}, e.g., accelerations, are solely computed from the known values at t^n. A stable time step must be smaller than a dilatational (i.e., pressure) wave moving through the model. The appropriate step-size can be described by

$$\Delta t = \frac{l_e}{c_d} \,, \tag{4.1}$$

with the characteristic length l_e of the smallest element and the dilatational wave speed $c_d = \sqrt{(\lambda + 2\mu)/\varrho}$ (see, e.g., Wriggers, 2008). With the material parameters of aluminum ($E = 70$ GPa, $\nu = 0.3$ and $\varrho = 2700$ kg/m^3), the dilatational wave speed is $c_d \approx 5907.31\,\mathrm{m/s}$. For instance, for a characteristic element length of $l_e = 1$ mm, the time step for the simulations needs to be lower than $\Delta t = 1.69 \cdot 10^{-7}$ s. The integration of the constitutive equations at the integration point level is done with an explicit Euler scheme. The stresses and accumulated plastic slip are updated using the relations

$$\boldsymbol{\sigma}^{n+1} = \boldsymbol{\sigma}^n + \Delta\boldsymbol{\sigma} \,, \tag{4.2}$$

$$\gamma^{n+1} = \gamma^n + \Delta\gamma \,, \tag{4.3}$$

respectively, with the stress increment evaluated by the incremental Hooke's from eq. (3.7)

$$\Delta\boldsymbol{\sigma} = \mathbb{C}\left[\Delta\varepsilon - \Delta\varepsilon_{\mathrm{p}}^{n}\right] , \tag{4.4}$$

the plastic strain increment

$$\Delta\varepsilon_{\mathrm{p}}^{n} = \sum_{\alpha=1}^{N} \Delta\gamma_{\alpha}^{n}\,\mathrm{sym}(\boldsymbol{M}_{\alpha}^{n}) , \tag{4.5}$$

the accumulated plastic slip increment

$$\Delta\gamma = \sum_{\alpha}^{N} |\Delta\gamma_{\alpha}^{n}| , \tag{4.6}$$

the plastic slip increment at t^n of component α

$$\Delta\gamma_{\alpha}^{n} = \Delta t\dot{\gamma}_{0}\mathrm{sgn}\left(\tau_{\alpha}^{n}\right)\left\langle \frac{|\tau_{\alpha}^{n}| - \tau_{n}^{C}}{\tau^{D}} \right\rangle^{m} , \tag{4.7}$$

the resolved shear stress at t^n of component α

$$\tau_{\alpha}^{n} = \boldsymbol{\sigma}^{n} \cdot \mathrm{sym}(\boldsymbol{M}_{\alpha}^{n}) , \tag{4.8}$$

the critical resolved shear stress at t^n

$$\tau_{n}^{C} = \tau_{0}^{C} + \Theta_{\infty}\gamma^{n} + \left(\tau_{\infty}^{C} - \tau_{0}^{C}\right)\left(1 - \exp\left(-\frac{\Theta_{0} - \Theta_{\infty}}{\tau_{\infty}^{C} - \tau_{0}^{C}}\gamma^{n}\right)\right) , \tag{4.9}$$

and the strain increment given in (3.9). Furthermore, with $\Delta\boldsymbol{Q}$ from eq. (3.14) the lattice rotation as an orthogonal tensor is updated with

$$\boldsymbol{Q}^{n+1} = \Delta\boldsymbol{Q}\boldsymbol{Q}^{n} . \tag{4.10}$$

4.2.2 Geometrically non-linear formulation: implicit time integration

With the evolution equations (3.19) and (3.21), $\boldsymbol{F}_\mathrm{e}$ and γ are integrated with an implicit Euler scheme (assuming the flow stress does not change during the time step), respectively:

$$\boldsymbol{F}_\mathrm{e}^{\,n+1} = \boldsymbol{F}_\mathrm{e}^{\,n} + \Delta t \dot{\boldsymbol{F}}_\mathrm{e}^{\,n+1} \quad , \quad \gamma^{n+1} = \gamma^n + \Delta t \dot{\gamma}^{n+1} \,. \tag{4.11}$$

Through the definition of the functions

$$\boldsymbol{G} = \boldsymbol{F}_\mathrm{e}^{\,n+1} - \boldsymbol{F}_\mathrm{e}^{\,n} - \Delta t \dot{\boldsymbol{F}}_\mathrm{e}^{\,n+1} \quad , \quad g = \gamma^{n+1} - \gamma^n - \Delta t \dot{\gamma}^{n+1} \,, \tag{4.12}$$

the unknowns $\boldsymbol{F}_\mathrm{e}^{\,n+1}$ and γ^{n+1} can be determined by solving the non-linear equations $\boldsymbol{G} = \boldsymbol{0}$ and $g = 0$ with the Newton-Raphson method. In a large deformations setting of an implicit FE analysis, the algorithmic tangent operator $\mathbb{C}^\mathrm{alg}$ needs to be evaluated. In ABAQUS Standard, the definition is as follows:

$$\mathbb{C}^\mathrm{alg} = \frac{1}{J \Delta t} \frac{\partial \boldsymbol{\tau}^{n+1}}{\partial \boldsymbol{D}^{n+1}} \,. \tag{4.13}$$

Hence,

$$\frac{\partial \boldsymbol{\tau}^{n+1}}{\partial \boldsymbol{D}^{n+1}} = \frac{\partial \boldsymbol{\tau}^{n+1}}{\partial \boldsymbol{F}_\mathrm{e}^{\,n+1}} \frac{\partial \boldsymbol{F}_\mathrm{e}^{\,n+1}}{\partial \boldsymbol{L}^{n+1}} \frac{\partial \boldsymbol{L}^{n+1}}{\partial \boldsymbol{D}^{n+1}} \,, \tag{4.14}$$

with a constant velocity gradient $\boldsymbol{L}^{n+1} = \boldsymbol{L}^n = \boldsymbol{L} = \text{const}$. The partial derivatives of eq. (4.14) can be found in Appendix C.

4.3 Application of routines: simple shear of a single crystal

The single crystal plasticity model introduced in Section 3.2 and 3.3 are implemented in the explicit VUMAT and implicit UMAT setting, respectively.

As a benchmark example, the finite element implementation is applied to a simple shear deformation of a crystal in standard orientation with the material parameters from Table 4.1. The values of the stress components $\sigma_{11}, \sigma_{22}, \sigma_{12}$ are plotted over the shear deformation (see Fig. 4.1). For the single crystal in initial orientation $\varphi_1 = 0,\ \Phi = 0,\ \varphi_2 = 0$, there is a deformation induced rotation of the crystal to $\varphi_1 = 0,\ \Phi = \pi/2,\ \varphi_2 = 0$ in both implicit and explicit cases.

Whereas the results for the shear stress σ_{12} are almost identical ($\leq 0.4\%$) for the explicit and implicit solution, there is a small difference in the normal stress components σ_{11} and σ_{22} ($\leq 2\%$). Overall, the results from the explicit and implicit solutions are in good agreement and the incremental implementation is verified.

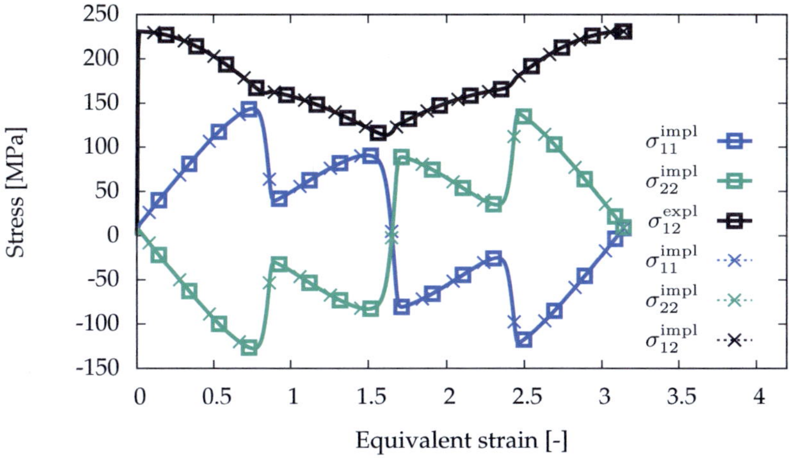

Figure 4.1: Simple shear: Comparison of implicit (solid line) and explicit (dashed line) solution for stress components vs strain diagram

Table 4.1: Material parameters for single crystal model

E [MPa]	ν [$-$]	$\dot{\gamma}_0$ [s^{-1}]	τ_0^C [MPa]	τ_∞^C [MPa]	τ_D [MPa]	m [$-$]	Θ_0 [MPa]	Θ_∞ [MPa]
70000	0.3	0.001	100	0	20	20	0	0

Chapter 5

Homogenization of the polycrystal behavior

5.1 Effective kinematic and dynamic properties

Heterogeneous material behavior at the microstructural level is averaged to predict the response of an effective medium at the macroscopic level, a procedure known as homogenization. The concept of a representative volume element (RVE) was introduced by Hill (1963). Therein, the RVE is located at every material point of a macroscopic specimen. Furthermore, the RVE must be small enough compared to the overall size of the specimen, and, simultaneously, sufficiently large to be representative of the microstructure of the specimen. Then the mechanical behavior of a specimen can be described using a RVE. Since it is assumed that the microstructure does neither contain pores nor cracks, the effective stresses and strains can be defined as the volume averages of the fluctuating micro-stresses and strains (Hill, 1963; 1967),

$$\bar{\boldsymbol{\sigma}} := \langle \boldsymbol{\sigma} \rangle = \frac{1}{V} \int\limits_{V_{\mathrm{RVE}}} \boldsymbol{\sigma}(\boldsymbol{x}) \, \mathrm{d}V \,, \qquad \bar{\boldsymbol{\varepsilon}} := \langle \boldsymbol{\varepsilon} \rangle = \frac{1}{V} \int\limits_{V_{\mathrm{RVE}}} \boldsymbol{\varepsilon}(\boldsymbol{x}) \, \mathrm{d}V \,, \tag{5.1}$$

respectively. The volume averages of stress and strain increments, $\langle \Delta\boldsymbol{\sigma} \rangle$ and $\langle \Delta\boldsymbol{\varepsilon} \rangle$, are calculated analogously to eq. (5.1). The Hill theorem (Hill, 1963) states that the macroscopic stress power is equal to the average of the microscopic stress power

$$\langle \boldsymbol{\sigma} \cdot \dot{\boldsymbol{\varepsilon}} \rangle = \langle \boldsymbol{\sigma} \rangle \cdot \langle \dot{\boldsymbol{\varepsilon}} \rangle = \bar{\boldsymbol{\sigma}} \cdot \dot{\bar{\boldsymbol{\varepsilon}}} \ . \tag{5.2}$$

This statement is valid for all kinematically admissible strain fields and all statistically admissible stress fields. Hence, it is independent of the material behavior of the constituents. The average stress and strain in a phase α can be determined as follows:

$$\boldsymbol{\varepsilon}_\alpha = \frac{1}{V_\alpha} \int \boldsymbol{\varepsilon} \, \mathrm{d}V \ , \qquad \boldsymbol{\sigma}_\alpha = \frac{1}{V_\alpha} \int \boldsymbol{\sigma} \, \mathrm{d}V \ . \tag{5.3}$$

In the context of large deformations, the effective quantities for the first Piola-Kirchhoff tensor and deformation gradient are given by (see e.g., Nemat-Nasser, 1999)

$$\bar{\boldsymbol{S}}^{1PK} = \frac{1}{V} \int \boldsymbol{S}^{1PK} \, \mathrm{d}V \quad \text{and} \quad \bar{\boldsymbol{F}} = \frac{1}{V} \int \boldsymbol{F} \, \mathrm{d}V \ , \tag{5.4}$$

respectively, with $\boldsymbol{S}^{1PK} = J \boldsymbol{\sigma} \boldsymbol{F}^{-\mathsf{T}}$.

5.2 Non-linear Hashin-Shtrikman type homogenization

Piecewise constant polarizations in the volume V_α of grain α are assumed (Willis, 1977; 1981; Ponte Castañeda and Suquet, 1998)

$$\boldsymbol{p}(\boldsymbol{x}) = \boldsymbol{\sigma}(\boldsymbol{x}) - \mathbb{C}_0[\boldsymbol{\varepsilon}(\boldsymbol{x})] \ , \tag{5.5}$$

with

$$\boldsymbol{p}(\boldsymbol{x}) = \sum_{\alpha=1}^{N} \chi_\alpha(\boldsymbol{x}) \boldsymbol{p}_\alpha, \ \ \chi_\alpha(\boldsymbol{x}) = \begin{cases} 1 & \forall \boldsymbol{x} \in \boldsymbol{V}_\alpha \\ 0 & \text{otherwise} \end{cases} \ . \tag{5.6}$$

The linear Hashin-Shtrikman method is generalized in terms of allowing a non-linear stress response $\boldsymbol{\sigma}_\alpha = \boldsymbol{\sigma}_\alpha(\boldsymbol{\varepsilon}_\alpha)$ in grain α. Furthermore, the stiffness $\mathbb{C}_0$ of the reference medium is a parameter, which takes into account statistics of strain and stress fields (Jöchen and Böhlke, 2012; Jöchen, 2013).

The polarization tensor $\mathbb{P}_0$

$$\mathbb{P}_0(\mathbb{C}_0, \boldsymbol{A}) = \frac{1}{4\pi \det(\boldsymbol{A})} \int_{||\boldsymbol{n}||=1} \mathbb{H}(\mathbb{C}_0, \boldsymbol{n}) \left(\boldsymbol{n} \cdot (\boldsymbol{A}^{-\mathsf{T}} \boldsymbol{A}^{-1} \boldsymbol{n}) \right)^{-3/2} \mathrm{d}n \,, \tag{5.7}$$

with

$$\mathbb{H} = \mathbb{I}^S \left(\boldsymbol{N} \square (\boldsymbol{n} \otimes \boldsymbol{n}) \right) \mathbb{I}^S, \quad \boldsymbol{N} = \boldsymbol{K}^{-1}, \quad \boldsymbol{K} = \mathbb{C}_0[[\boldsymbol{n} \otimes \boldsymbol{n}]], \tag{5.8}$$

relates the stress polarizations and strain fluctuations to each other. In rate form, the interaction law reads

$$\Delta \boldsymbol{\sigma}_\alpha - \Delta \bar{\boldsymbol{\sigma}} = \mathbb{L}[\Delta \boldsymbol{\varepsilon}_\alpha - \Delta \bar{\boldsymbol{\varepsilon}}], \qquad \mathbb{L} = \mathbb{C}_0 - \mathbb{P}_0^{-1} \,. \tag{5.9}$$

For a spherical inclusion in an isotropic material, the polarization tensor $\mathbb{P}_0$ simplifies to

$$\mathbb{P}_0 = \frac{1}{3K_0 + 4G_0} \mathbb{P}_1^{\mathrm{iso}} + \frac{3(K_0 + 2G_0)}{5G_0(3K_0 + 4G_0)} \mathbb{P}_2^{\mathrm{iso}} \,, \tag{5.10}$$

with the isotropic projectors $\mathbb{P}_1^{\mathrm{iso}} = 1/3(\boldsymbol{I} \otimes \boldsymbol{I})$ and $\mathbb{P}_2^{\mathrm{iso}} = \mathbb{I}^S - \mathbb{P}_1^{\mathrm{iso}}$.

For the special case of an isotropic material stiffness and isotropic medium stiffness ($\mathbb{C}$ and $\mathbb{C}_0$ isotropic), and the difference between material and reference stiffness being proportional to the second isotropic projector ($\mathbb{C} - \mathbb{C}_0 \sim \mathbb{P}_2^{\mathrm{iso}}$), the following equations are derived for the stress

$$\Delta \boldsymbol{\sigma}^\circ = 3K \Delta \varepsilon^\circ \,, \tag{5.11}$$

$$\Delta \boldsymbol{\sigma}' = 2G \Big(\Delta \bar{\boldsymbol{\varepsilon}}' + \big((1 - \lambda^{\mathrm{HS}}) \Delta \boldsymbol{\varepsilon}_{\mathrm{p}} + \lambda^{\mathrm{HS}} \langle \Delta \boldsymbol{\varepsilon}_{\mathrm{p}} \rangle \big) \Big) \,. \tag{5.12}$$

In this formulation, the scalar homogenization parameter $\lambda^{\mathrm{HS}} \in [0,1]$ controls the stiffness response. Considering eq. (5.12), the upper and lower homogenization bounds can be described with an appropriate choice for λ^{HS}. Then, the Voigt solution with homogeneous strains for $\lambda^{\mathrm{HS}} = 0$

$$\Delta\boldsymbol{\sigma}' = 2G\big(\Delta\bar{\varepsilon}' - \Delta\varepsilon_{\mathrm{p}}\big) \tag{5.13}$$

and the Reuss solution with homogeneous stresses for $\lambda^{\mathrm{HS}} = 1$

$$\Delta\boldsymbol{\sigma}' = 2G\big(\Delta\bar{\varepsilon}' - \langle\Delta\varepsilon_{\mathrm{p}}\rangle\big) = \langle\Delta\boldsymbol{\sigma}'\rangle\ . \tag{5.14}$$

can be obtained as a special case, respectively.
A proof of the corresponding derivations can be found in Appendix A. The sensitivity of the homogenization parameter λ^{HS} is discussed in detail in Section 7.5.

5.3 Distribution of the components of the stress tensor $\boldsymbol{\sigma}$

The influence of the aforementioned homogenization procedure in a plane strain compression (PSC) test is investigated. Therefore, an initially quasi-isotropic texture, generated with 1000 random orientations, is subjected to 2% and 50% thickness reduction in the PSC simulation. The influence of the homogenization parameter λ^{HS} on the stress distribution of the stress components for 2% thickness reduction is shown in Tab. 5.1, where for each stress component and λ^{HS} combination, a histogram is plotted with a superimposed best-fit normal distribution.
In the Reuss-Sachs case with $\lambda^{\mathrm{HS}} = 1$, the stresses in all phases are equal, i.e., a single peak is visible in the stress-distribution diagram. With decreasing λ^{HS}, the stress response from the different phases is diversified with a trend towards a pronounced Gaussian distribution. Furthermore,

in the Taylor case with $\lambda^{\mathrm{HS}} = 0$, all stress components exhibit a normal distribution (Gaussian distribution).
In the case of a larger PSC deformation, as it, e.g., can be seen for 50% thickness reduction in Tab. 5.2, a typical rolling texture is obtained and preferred crystal orientations are present. Due to this, only the shear stresses σ_{12} and σ_{23} show a Gaussian distribution, whereas the other stress components show distributions with distinct stress level peaks based on the preferred orientations.
The numerical findings confirm as shown the theoretical result that for soft comparison materials, the stress distribution is close to the Reuss-Sachs approximation, whereas for very stiff comparison materials, the distribution of strain increments or strain rates is close to the classical Voigt-Taylor approximation.
Therefore, the suggested homogenization scheme represents a mean field approach with variable statistical properties of the stress and strain fields. As mentioned before, the appropriate comparison material is determined based on a suitable texture approximation which depends itself on the stress fluctuations.

Table 5.1: Stress distribution for initially quasi-isotropic texture and 2% thickness reduction

λ^{HS}	σ_{11}	σ_{22}	σ_{33}	σ_{12}	σ_{13}	σ_{23}
1.0	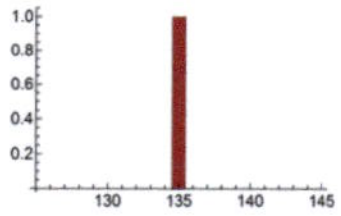	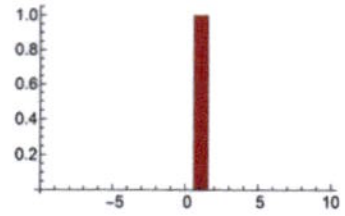	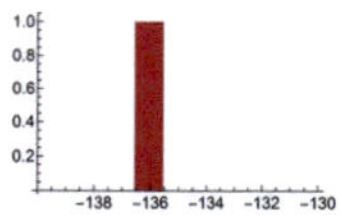	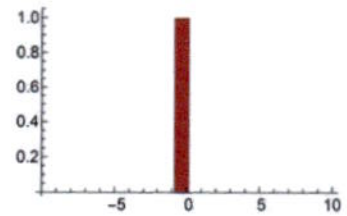	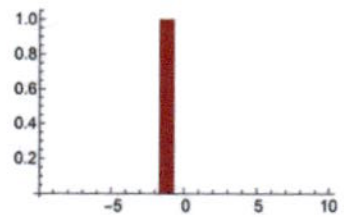	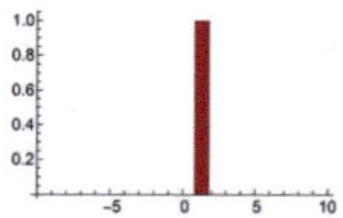
0.99	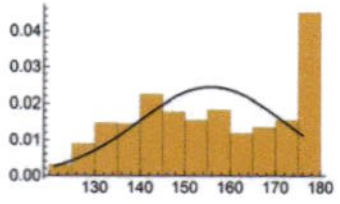	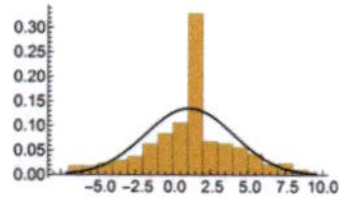	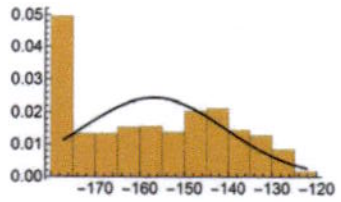	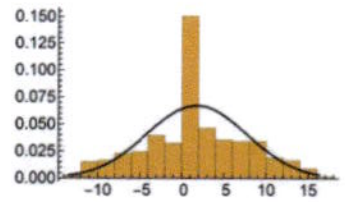	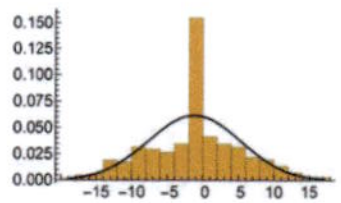	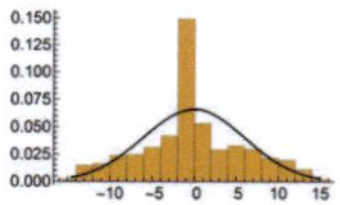
0.97	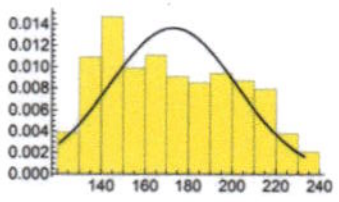	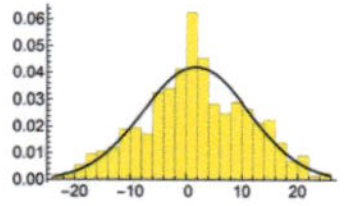	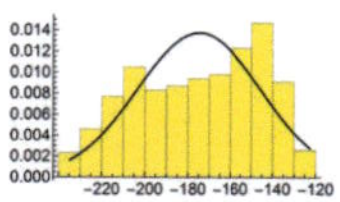	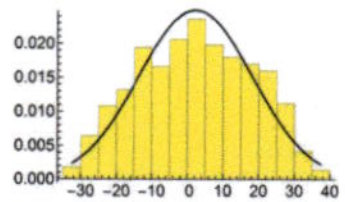	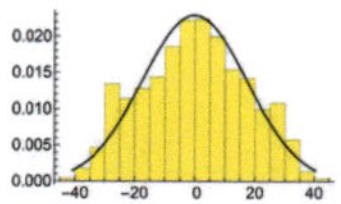	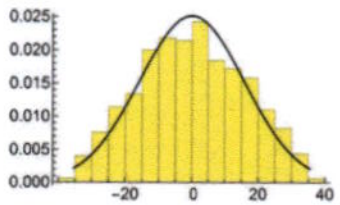
0.5	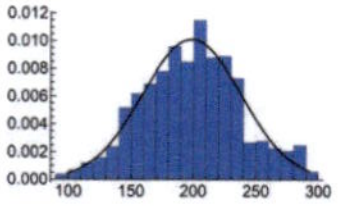	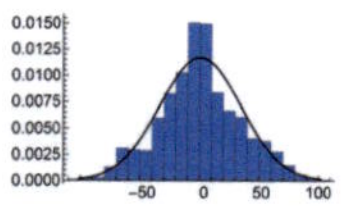	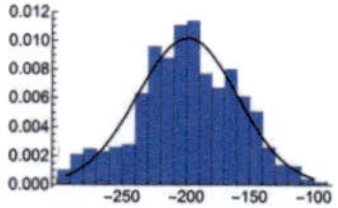	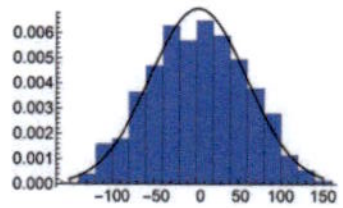	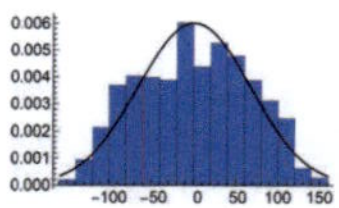	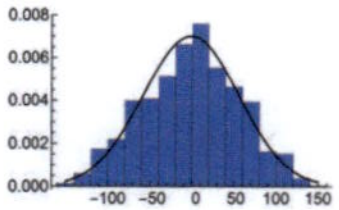
0.0	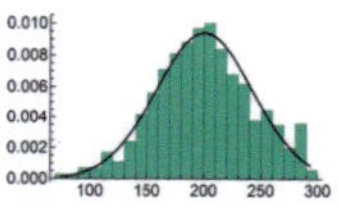	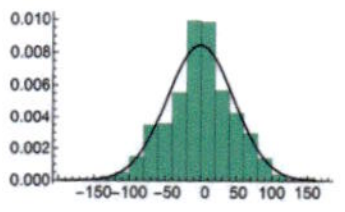	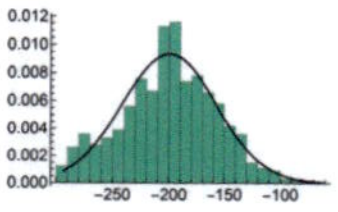	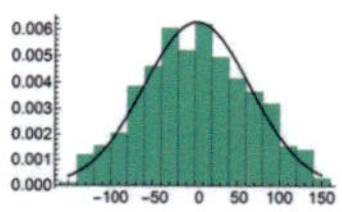	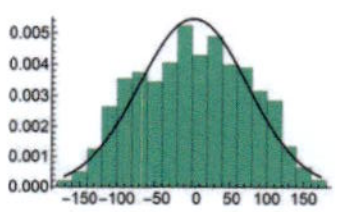	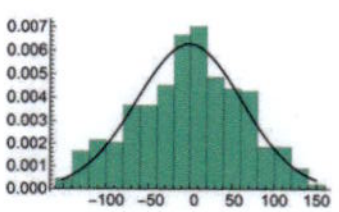

Table 5.2: Stress distribution for initially quasi isotropic-texture and 50% thickness reduction

λ^{HS}	σ_{11}	σ_{22}	σ_{33}	σ_{12}	σ_{13}	σ_{23}
0.9995	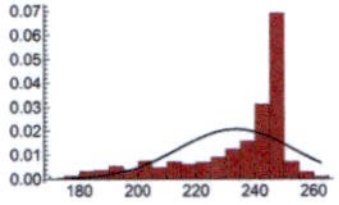	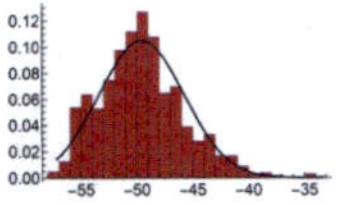	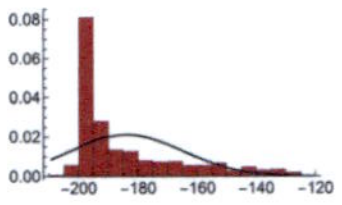	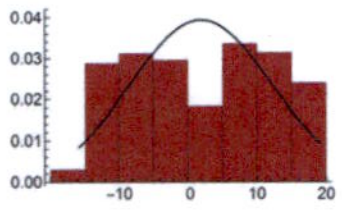	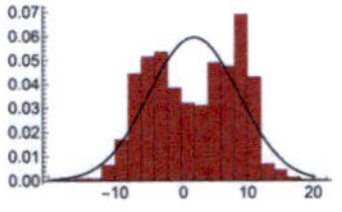	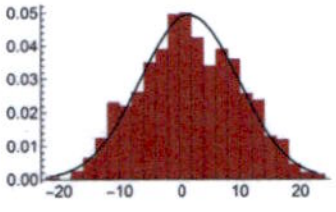
0.99	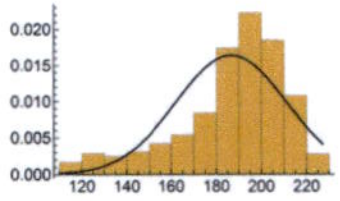	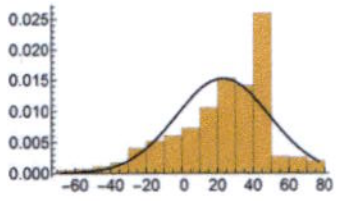	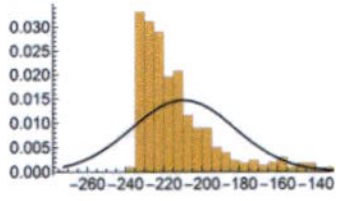	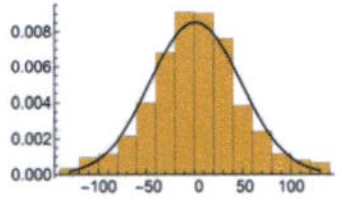	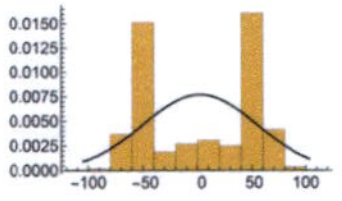	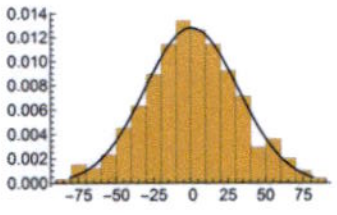
0.97	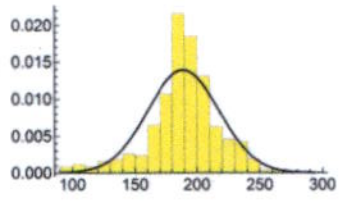	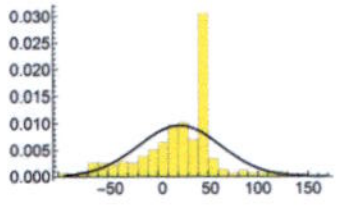	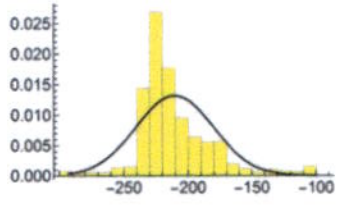	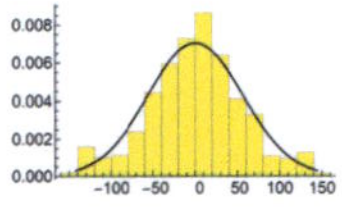	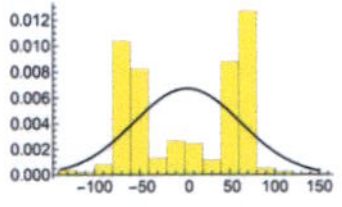	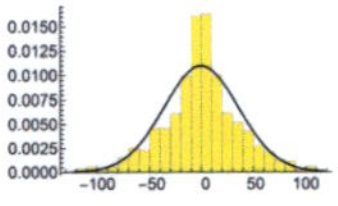
0.5	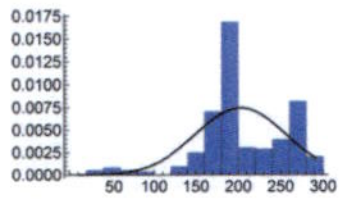	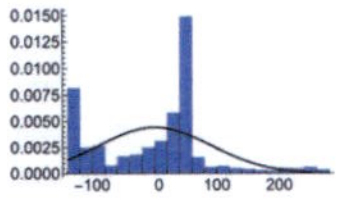	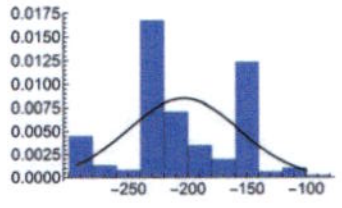	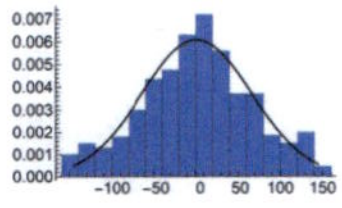	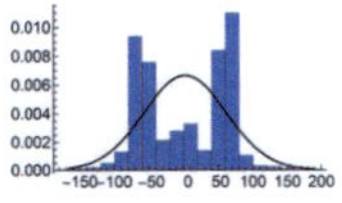	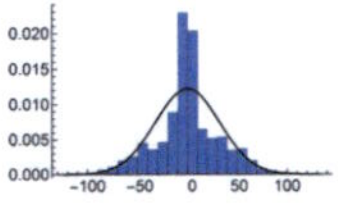
0.0	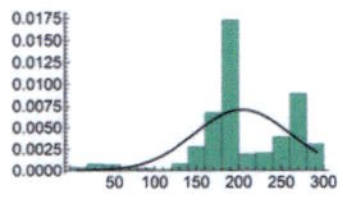	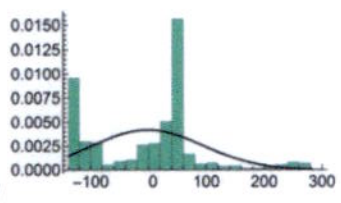	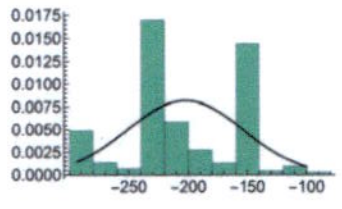	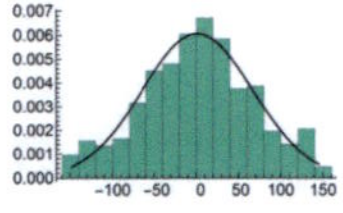	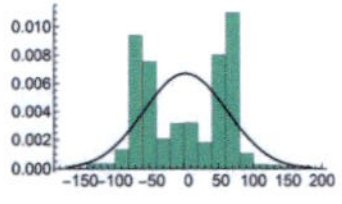	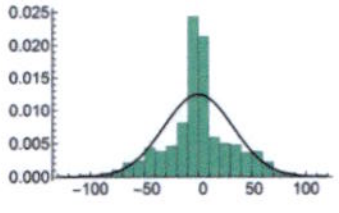

Chapter 6

Experimental characterization of aluminum EN AW-3104

6.1 Preliminaries

In the following, experimental results are shown for the EN AW-3104 alloy (for the chemical composition of the alloy see Tab. 6.1). The deep drawing of packaging material of this or similiar alloys has already been studied in previous works (e.g., Hutchinson et al., 1989; Ren and Das, 1998; Hirsch, 2005). Here, tensile tests, hydraulic bulge tests, earing measurements, Nakajima tests as well as crystallographic texture measurements have been performed by Constellium to characterize the material behavior.

Table 6.1: EN AW-3104 (EN AW-Al Mn1Mg1Cu): Chemical composition of alloy with remainder Al in wt. % (DIN EN 573-3, 2013)

Chemical composition [weight % max. or range]								
Si	Fe	Cu	Mn	Mg	Zn	Ti	Ga	V
0.6	0.8	0.05 – 0.25	0.8 – 1.4	0.8 – 1.3	0.25	0.1	0.05	0.05

6.2 Flow curves

Uniaxial tension tests were performed at Constellium on the can body stock aluminum EN AW Al-3104 (Suisse Technology Partners AG Test Report M-13-005, 2013). Therein, the experiments were performed based on EN 10002-1 (2001). The sheet metal specimen had a thickness of $t = 0.264$ mm and a width of $w = 15$ mm. Furthermore, the experiments were performed at 10 $\mathrm{mm/s}$ testing speed.

Different orientations of specimens in the sheet metal were tested, starting with the rolling direction (RD), and then in $15°$ steps up to $90°$ (see Fig. 6.1–6.3), where 10 experiments were performed in each configuration.

In the tensile experiments, the EN AW-3104 aluminum alloy exhibits a Portevin-Le Chatelier (PLC) effect, which has been well documented before (e.g., Park and Morris, 1993) and there fails comparably early at a rather low overall strain.

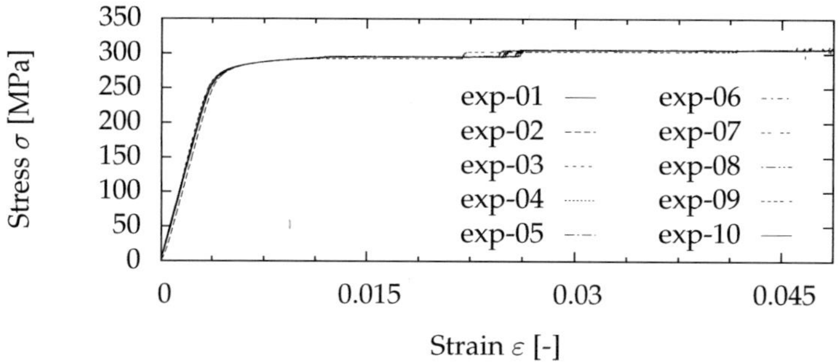

Figure 6.1: Tensile test in $0°$ to RD on 10 specimens (Suisse Technology Partners AG Test Report M-13-005, 2013)

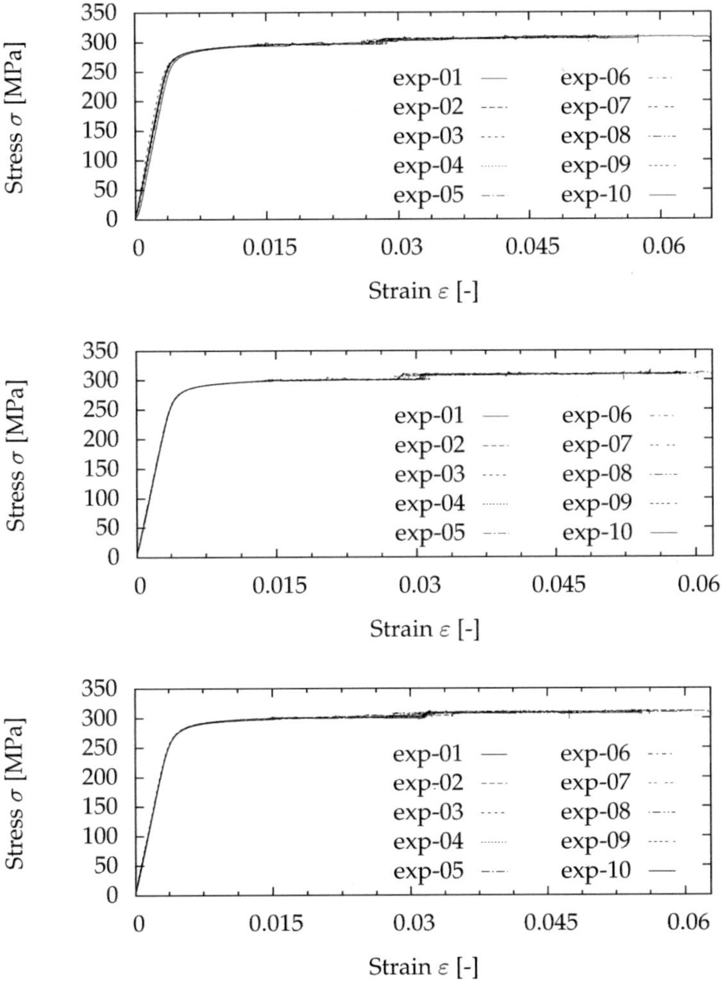

Figure 6.2: Tensile test in 15° (top), 30° (middle) and 45° (bottom) to RD on 10 specimens each (Suisse Technology Partners AG Test Report M-13-005, 2013)

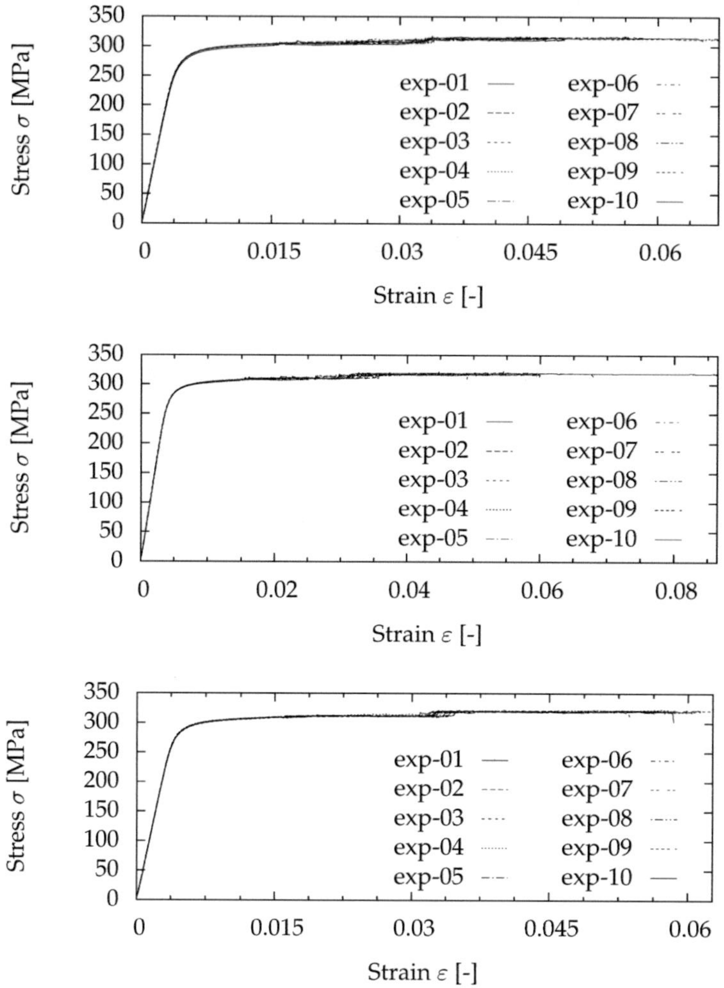

Figure 6.3: Tensile test in 60° (top), 75° (middle) and 90° (bottom) to RD on 10 specimens each (Suisse Technology Partners AG Test Report M-13-005, 2013)

6.3 Lankford coefficient

Anisotropy of sheet metals can be characterized with Lankford coefficients (r-values). To determine the r-value, specimens are cut from the sheet at different angles to the RD ($0°$, $15°$, $15°$, $30°$, $45°$, $60°$, $75°$, $90°$), and the strain ratio of width to thickness direction is calculated by $r = \varepsilon_w/\varepsilon_t$ (Lankford et al., 1950). All results shown here were measured by Constellium (Suisse Technology Partners AG Test Report M-13-005, 2013). The results are summarized in Fig. 6.4. Typically, Lankford coefficients are measured in $0°$, $45°$, $90°$ to the RD, in the literature. The additional measurements performed here give further insight into the anisotropic behavior of the alloy. Whereas the specimen behaves rather isotropic ($r \approx 1$) at $45°$ to the RD, the most anisotropic behavior is observed in the RD and the transverse direction (TD) ($r \neq 1$ at $0°$ and $90°$). While the measurements in RD are consistent, the scatter of the data is pronounced in TD.

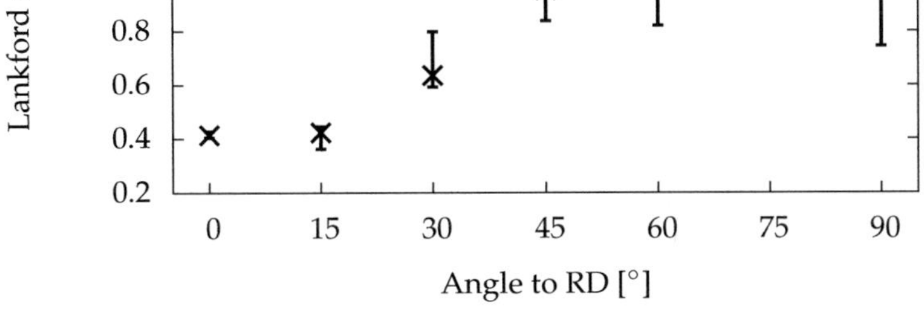

Figure 6.4: Lankford coefficients (r-values) (Suisse Technology Partners AG Test Report M-13-005, 2013)

6.4 Hydraulic Bulge tests

In a hydraulic bulge test, a sheet metal test specimen is placed over a circular hole, clamped and bulged outward by a fluid medium (Olsen, 1920; Hosford and Caddell, 2011). During the test, the pressure of the fluid, the radius of the bulge, and the strain field (using optical measurement systems) are measured. Furthermore, the die diameter is 100 mm and the piston speed is 0.6 mm/s. The results were measured by Constellium (Suisse Technology Partners AG Test Report E-13-025, 2013) with the method proposed in DIN EN ISO 16808 (2014).

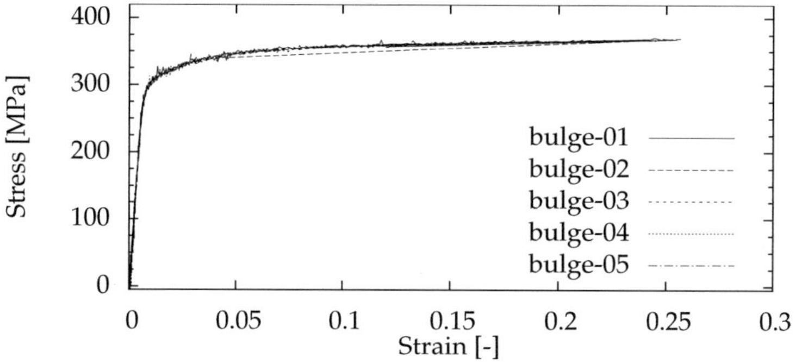

Figure 6.5: Hydraulic bulge test (Suisse Technology Partners AG Test Report E-13-025, 2013)

6.5 Earing profiles

Deep drawing of a circular sheet is performed at Constellium (Constellium CRV Test Report 43-11-1-3, 2013), and the non-uniform cup height (earing) due to the anisotropy of the sheet metal is measured (Banabic, 2010).

The single measurements around the circumference of the cup are plotted over the angle to the RD. For the BUP measurements (initial blank diameter: 64 mm), three data sets were measured which are shown as raw data (Fig. 6.6) and as symmetrized data (Fig. 6.7). Analogously, for the CUP measurements five data sets were measured which are shown as raw data (Fig. 6.8) and as symmetrized data (Fig. 6.9). The observed earing profile is typical for low-alloyed aluminum used in the packaging industry, exhibiting six ears.

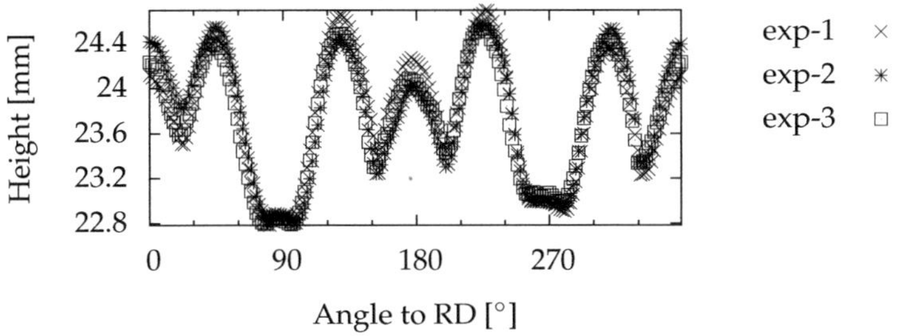

Figure 6.6: BUP: Experimentally measured earing profile (Constellium CRV Test Report 43-11-1-3, 2013)

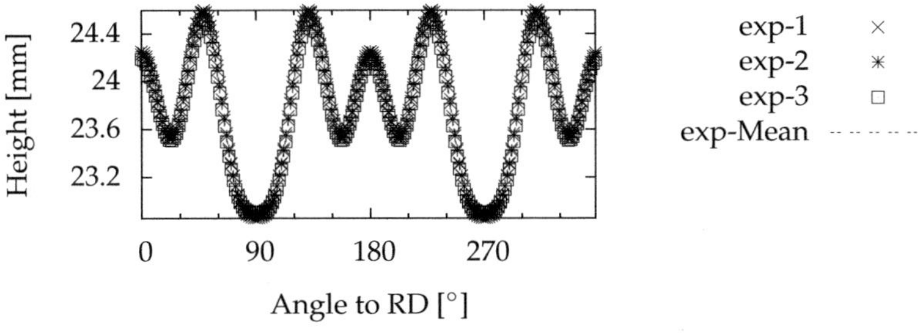

Figure 6.7: BUP: Experimentally measured symmetrized earing profile (Constellium CRV Test Report 43-11-1-3, 2013)

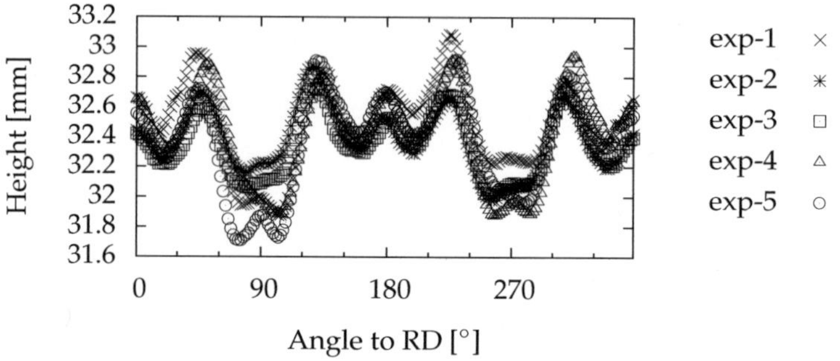

Figure 6.8: CUP: Experimentally measured earing profile (Constellium CRV Test Report 43-11-1-3, 2013)

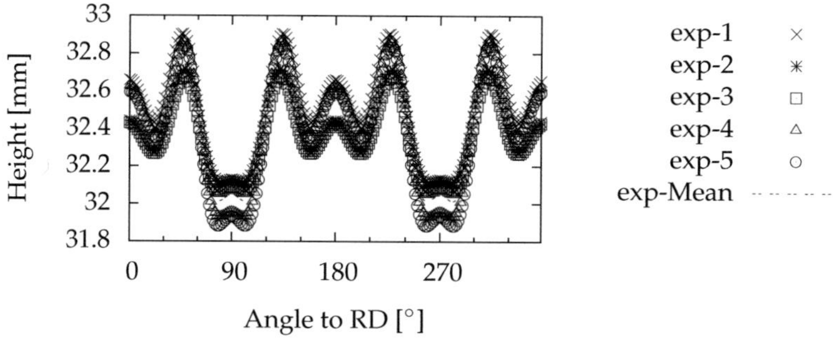

Figure 6.9: CUP: Experimentally measured symmetrized earing profile (Constellium CRV Test Report 43-11-1-3, 2013)

6.6 Forming limit curves

For the forming limit curves, biaxial Nakajima tests (Hosford and Caddell, 2011) were performed at Constellium (Suisse Technology Partners AG Test Report E-13-030, 2013) with a punch velocity of 1 $\mathrm{mm/s}$.

The specimens were orientated in rolling direction and a random pattern was applied to the sample in order to measure the strains with the ARAMIS system. Three geometries were tested and the results are summarized in Fig. 6.10. This is not a full FLC according to DIN EN ISO 12004-2 (2008) due to the fact that only three geometries have been tested.

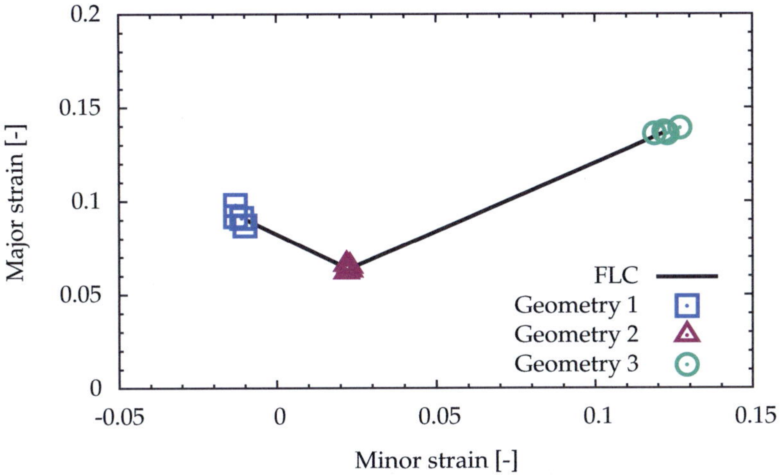

Figure 6.10: Forming limit curve (Suisse Technology Partners AG Test Report E-13-030, 2013)

6.7 Crystallographic texture measurements

The crystallographic texture measurements were performed at Constellium (Constellium CRV Test Report 49-14-1-5, 2013) with the X-ray diffraction technique (Kocks et al., 2000).

For every dataset measurement, orientation distribution function (ODF) and (100), (110), and (111) pole figures are shown. The sample geometries are 50x50 mm for the initial texture and 13x24 mm for the subsequent forming operations were used. The sections through Euler space with $\varphi_2 =$ const. of the CODF as well as (100), (110), and (111) pole figures are shown. For the texture after deformation, measurement in RD as well as in TD have been performed.
The initial texture is a typical rolling texture with a pronounced brass component in the $\varphi_2 = 0°$ section and a significant S component in the $\varphi_2 = 65°$ section of Fig. 6.11. For the texture after deep drawing in the small scale experiment (BUP) in RD, there is a shift in the brass and S component in the $\varphi_2 = 0°$ and $\varphi_2 = 65°$ sections of Fig. 6.12, respectively. For the same sample in TD (Fig. 6.13), the Goss component has vanished and only the shifted brass component is present for $\varphi_2 = 0°$. For the sample with original beverage can geometry, there have been measurements in different cup heights of the deep drawn specimen. The middle of the cup height as well as the top of the cup height for both RD and TD have been distinguished, respectively. Comparing the measurements in RD, the texture at the middle of the cup height in Fig. 6.14 is not as sharp as the texture at the top of the cup height in Fig. 6.16. As in the case of the small scale sample BUP, there is a shift in the brass and S component. Furthermore, for the measurements in TD, the texture at the top of the cup height in Fig. 6.17 is sharper, not only in the CODF but also in the pole figures, best visible in the (110) pole figure compared to the middle of cup height in Fig. 6.15.

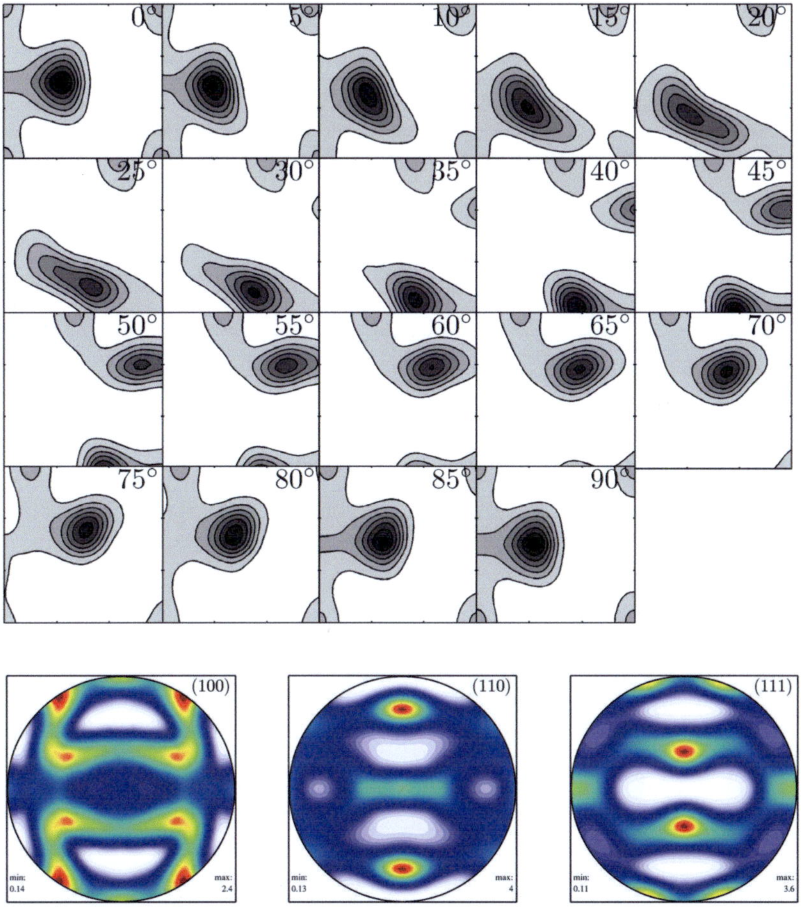

Figure 6.11: Experimentally measured texture of initial sheet (Constellium CRV Test Report 49-14-1-5, 2013): CODF sections with $\varphi_2 = $ const. with contours at 1, 2, 3, 4, 5 and 6 (top); (100), (110), and (111) pole figures (bottom, from left to right)

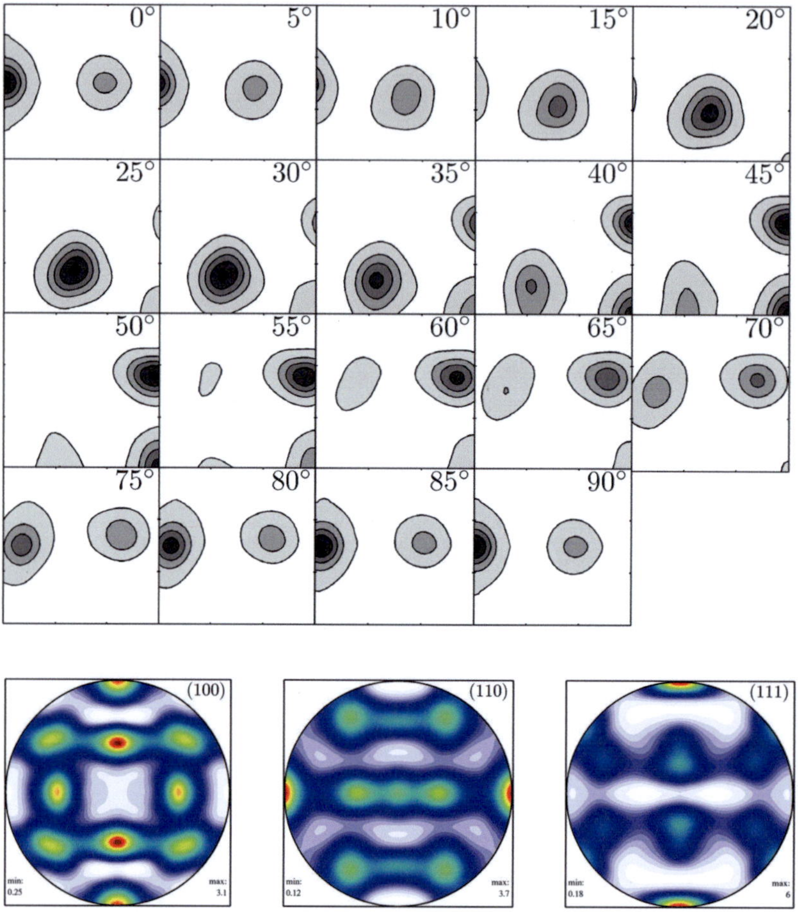

Figure 6.12: Experimentally measured texture of BUP in RD (Constellium CRV Test Report 49-14-1-5, 2013): CODF sections with $\varphi_2 = \text{const.}$ with contours at 1, 3, 5, 7 and 9 (top); (100), (110), and (111) pole figures (bottom, from left to right)

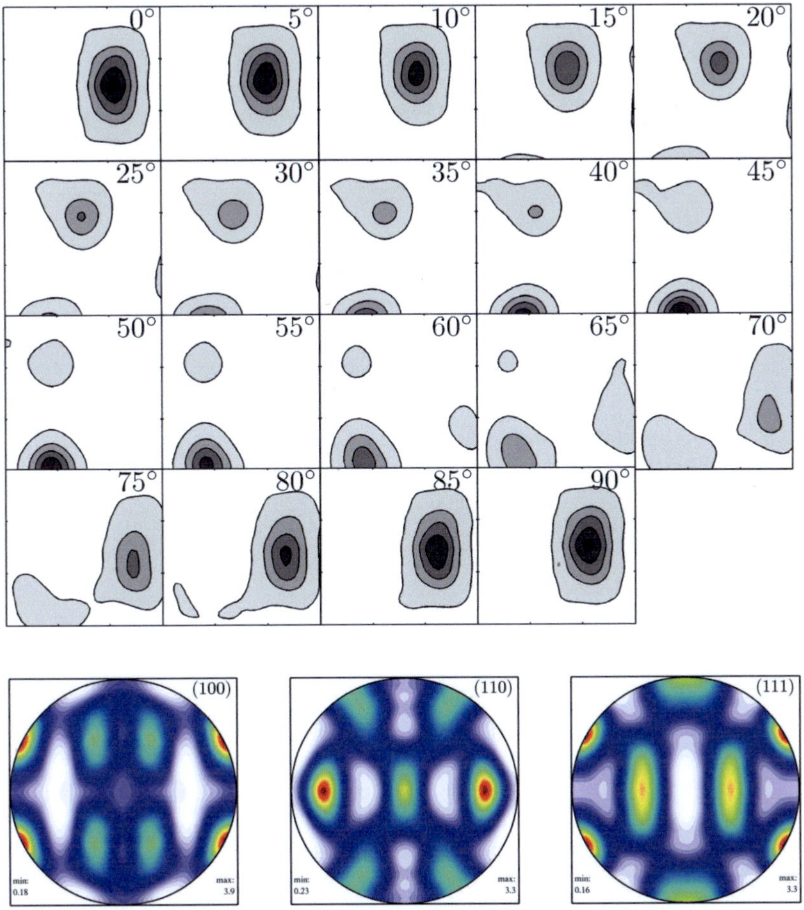

Figure 6.13: Experimentally measured texture BUP in TD (Constellium CRV Test Report 49-14-1-5, 2013): CODF sections with φ_2 = const. with contours at 1, 3, 5, 7 and 9 (top); (100), (110), and (111) pole figures (bottom, from left to right)

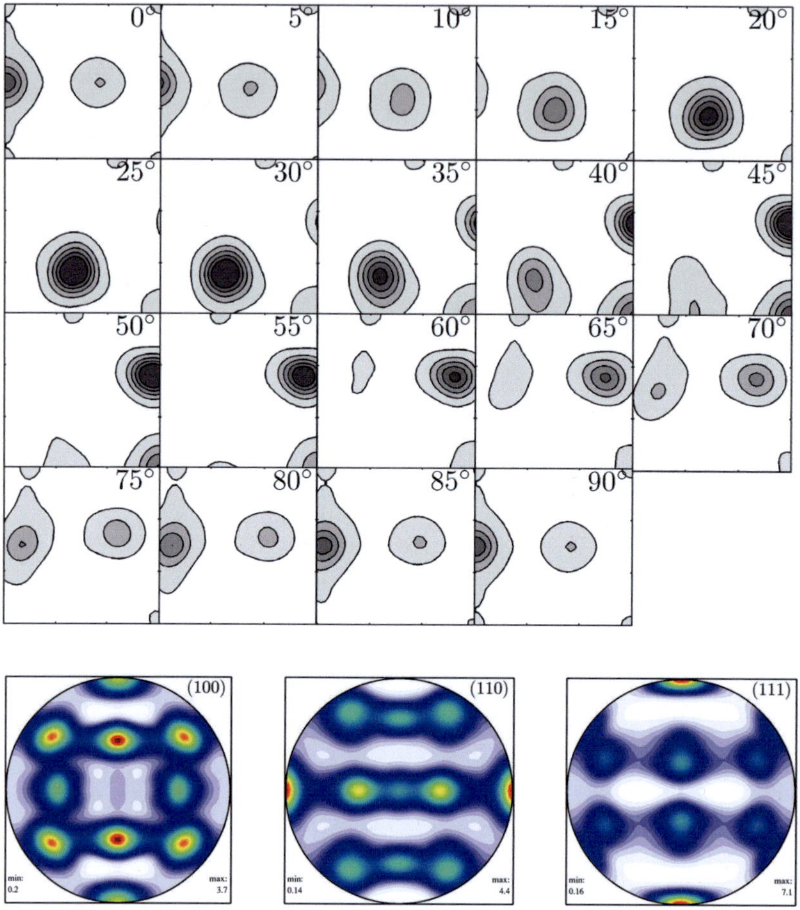

Figure 6.14: Experimentally measured texture middle of cup height in RD (Constellium CRV Test Report 49-14-1-5, 2013): CODF sections with $\varphi_2 = \text{const.}$ with contours at 1, 3, 5, 7 and 9 (top); (100), (110), and (111) pole figures (bottom, from left to right)

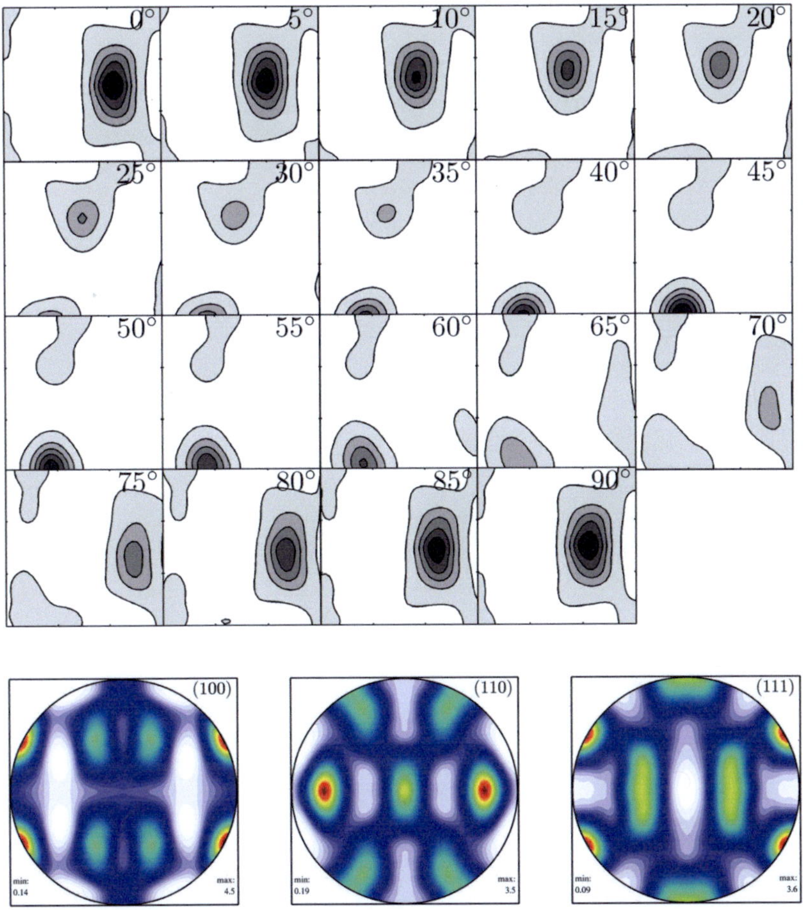

Figure 6.15: Experimentally measured texture middle of cup height in TD (Constellium CRV Test Report 49-14-1-5, 2013): CODF sections with $\varphi_2 = \text{const.}$ with contours at 1, 3, 5, 7, 9 and 11 (top); (100), (110), and (111) pole figures (bottom, from left to right)

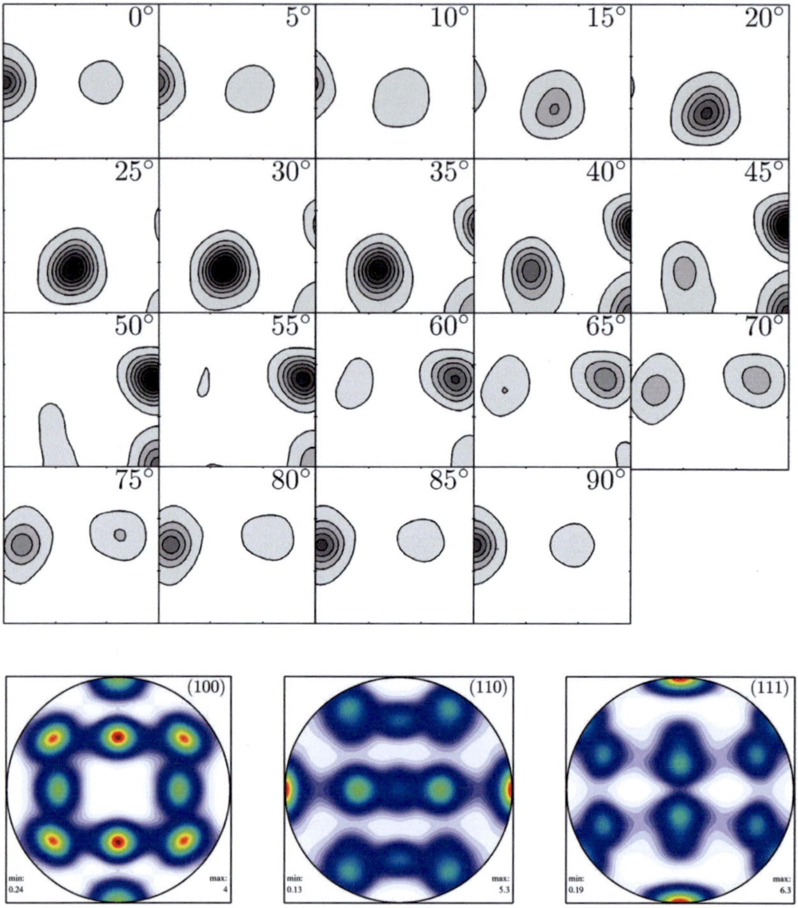

Figure 6.16: Experimentally measured texture top of cup height in RD (Constellium CRV Test Report 49-14-1-5, 2013): CODF sections with $\varphi_2 =$ const. with contours at 1, 3, 5, 7, 9, 11 and 13 (top); (100), (110), and (111) pole figures (bottom, from left to right)

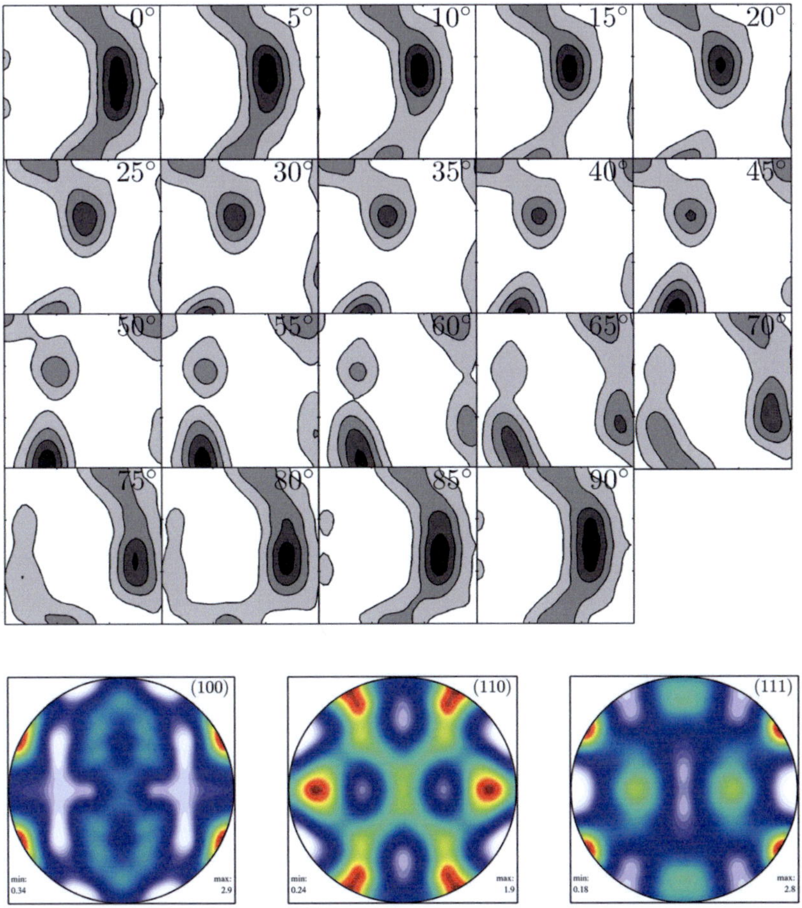

Figure 6.17: Experimentally measured texture top of cup height in TD (Constellium CRV Test Report 49-14-1-5, 2013): CODF sections with $\varphi_2 = $ const. with contours at 1, 2, 3, 4, and 5 (top); (100), (110), and (111) pole figures (bottom, from left to right) (Constellium CRV Test Report 49-14-1-5, 2013)

Chapter 7

Parameter identification and microstructure description

7.1 Material parameters of the crystal plasticity constitutive law

The elastic material parameters are taken for an aluminum alloy with the Young's modulus and Poisson's ration equal to $E = 70000$ MPa and $\nu = 0.3$, respectively. With the texture measurements of an undeformed sheet, a uniaxial tension test simulation is performed and the stress result is compared to the experimental results. The measured orientations serve as an input at the Gauss point level for the finite element simulation, where a Taylor type ($\lambda^{\mathrm{HS}}_{\mathrm{PI,I}} = 0$) and a Reuss type ($\lambda^{\mathrm{HS}}_{\mathrm{PI,II}} = 0.999$) homogenization is performed in order to simulate the polycrystalline behavior of the sample. Both of the identified sets of material parameters are shown in Tab. 7.1.

Table 7.1: Material parameters for the hardening behavior: parameter identification with Taylor type $\lambda^{\mathrm{HS}}_{\mathrm{PI,I}} = 0$) or Reuss type ($\lambda^{\mathrm{HS}}_{\mathrm{PI,II}} = 0.999$) homogenization parameter

PI	$\lambda^{\mathrm{HS}}_{\mathrm{PI}}$ [–]	$\dot{\gamma}_0$ [s^{-1}]	τ_0^C [MPa]	τ_∞^C [MPa]	τ_D [MPa]	m [–]	Θ_0 [MPa]	Θ_∞ [MPa]
I	0.0	0.001	83	88	20	20	1	2
II	0.999	0.001	93	115	20	20	8000	0

These material parameters are used for the subsequent deep drawing and localization behavior simulations. A tensile test with the different set of material parameters in comparison to the experiment is shown in Fig. 7.1.

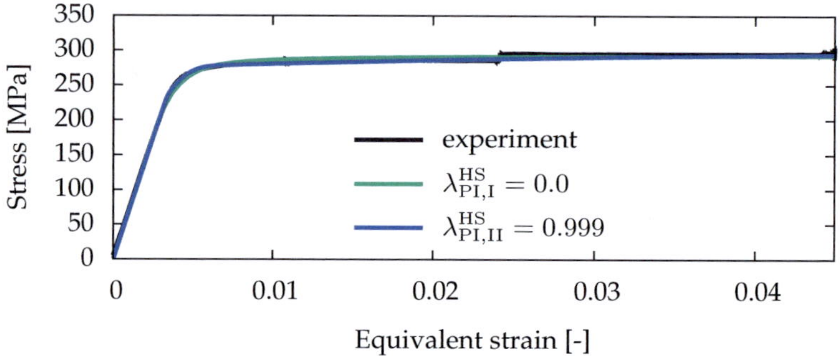

Figure 7.1: Tension test: Experiment and simulation with identified material parameters

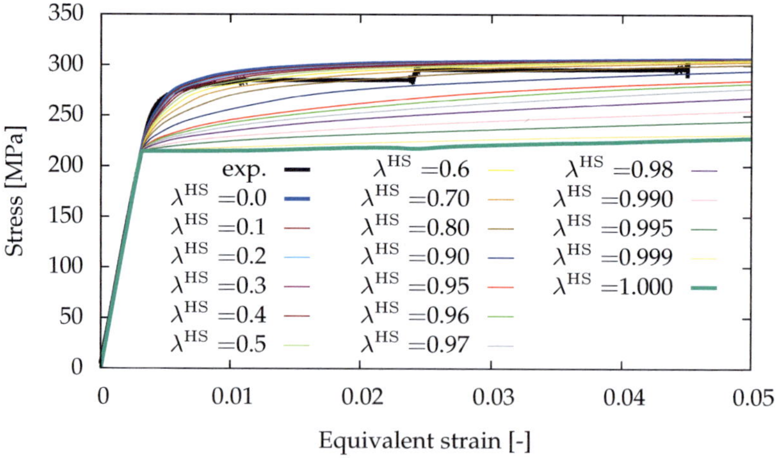

Figure 7.2: Influence of λ^{HS} on stress-strain curve in tensile test (parameter identification with $\lambda^{HS}_{PI,I} = 0$)

In Fig. 7.2 and 7.3 the influence of the homogenization parameter λ^{HS} is shown for a tensile test simulation and the resulting stress strain curve. The Taylor solution ($\lambda^{\mathrm{HS}} = 0$) leads to the highest stress level (upper bound), while the Reuss solution ($\lambda^{\mathrm{HS}} = 1$) has the lowest stress level (lower bound). The homogenization parameter is sensitive to small changes in the vicinity of the Reuss bound which lead to considerable changes in the stress regime. This behavior is described in more detail in Section 7.3.

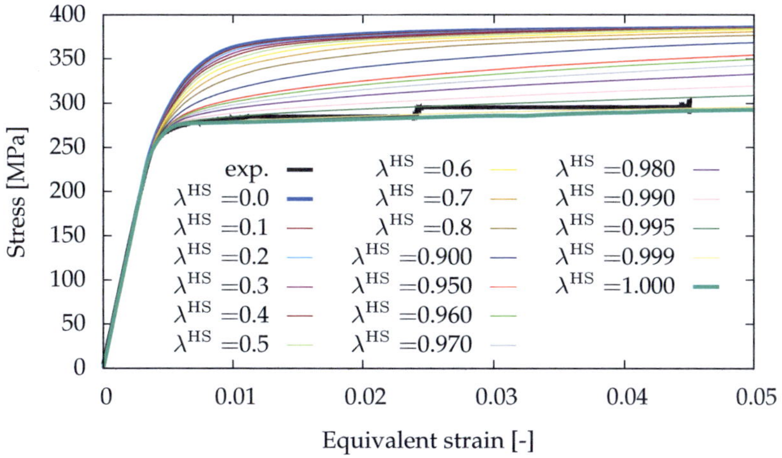

Figure 7.3: Influence of λ^{HS} on stress-strain curve in tensile test (parameter identification with $\lambda^{\mathrm{HS}}_{\mathrm{PI,II}} = 0.999$)

7.2 Low dimensional description of the crystallographic texture

The texture information is taken into account in the constitutive modeling in terms of orientation information which is used in the crystal plasticity model. As the number of experimentally measured orientations is significant (about 13500 orientations per specimen) and a homogenization is performed at the Gauss point level, the number of orientations needs to be reduced significantly in order to perform simulations in a feasible amount of time.

The method used in this work was proposed by Jöchen and Böhlke (2013) and briefly summarized in Section 3.5. Therein, all available orientations in Euler space are transformed into the fundamental zone with crystal and specimen symmetry transformations. The fundamental zone is subdivided into a prescribed number of boxes. All orientations in one box are averaged and, per box, an average orientation and a weighting factor of this average orientation is exported to establish a reduced dataset. By varying the size of the boxes, a variable approximation of the higher dimensional crystallographic texture, based on a low dimensional discrete set of orientations, can be obtained. With this approach, several reductions of the original orientation dataset are generated. They are shown in Tab. 7.2, where pole figures and CODFs are compared to the initial texture.

7.3 Influence of homogenization parameter λ^{HS} on deformation texture

To study the influence of the homogenization scheme on the texture development during deformation, plane strain compression (PSC) simulations are performed for several exemplary cases. In the first case, a quasi-isotropic texture is created with the help of random orientations. Therein, each orientation is created with a set of random numbers x_1, x_2, x_3 with $x_i \in [0, 1]$, $i = 1, 2, 3$, and their corresponding Euler angles with the relation (see, e.g., Murnaghan, 1962; Bunge, 1982; Böhlke, 2000):

$$\varphi_1 = 2\pi x_1 \ , \quad \Phi = \arccos\left(2x_2 - 1\right) \ , \quad \varphi_2 = 2\pi x_3 \ . \tag{7.1}$$

Table 7.2: Pole figures (100 and 111) and CODF of initial and reduced aluminum alloy EN AW-3104 orientation data

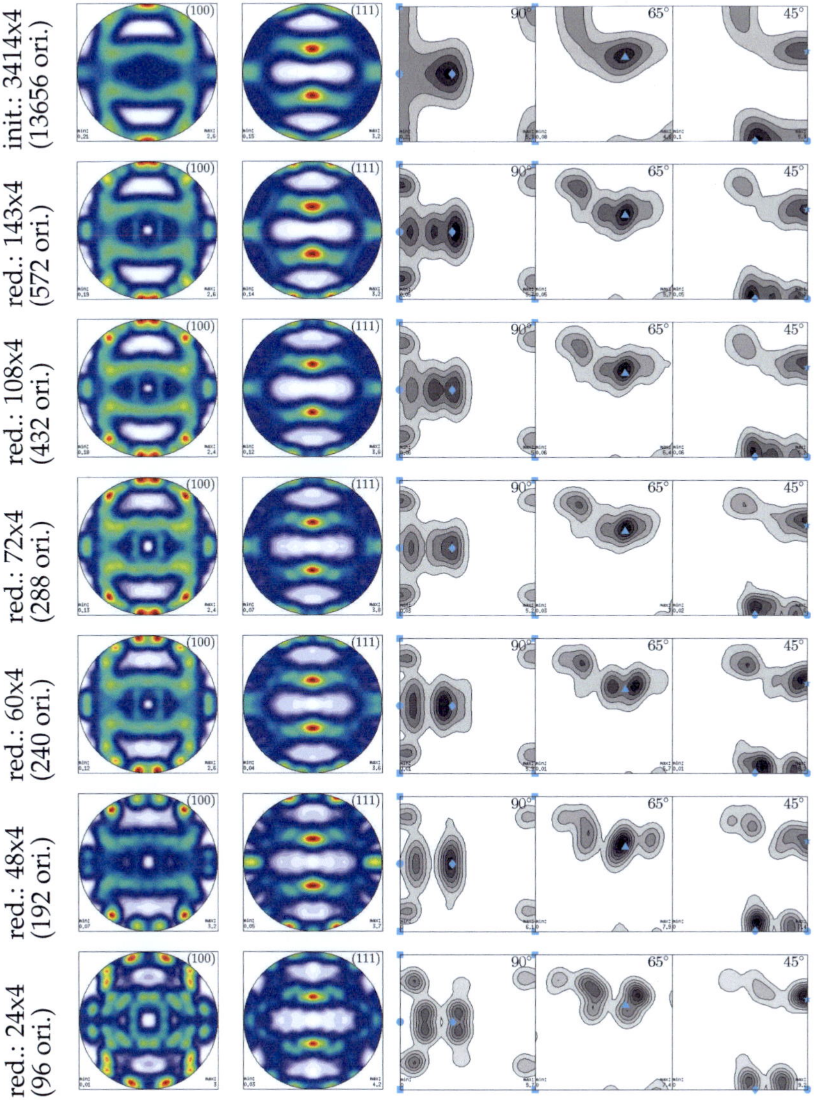

In Tab. 7.3, the results of a quasi-isoropic initial texture with 1000 orientations and the rolling texture after 50% thickness reduction are summarized in form of CODF plots (sections with $\phi_2 = 90°$, $\phi_2 = 65°$ and $\phi_2 = 45°$). Two simulations are shown which cover the domain of λ^{HS}. For $\lambda^{\mathrm{HS}} = 0.99$, there is a pronounced Brass, Goss and S component and no cube components in the $\phi_2 = 90°$-plot. Furthermore, for $\lambda^{\mathrm{HS}} = 0$, there exist cube components and there is a split into two peaks in the vicinity of the Brass component. The same PSC simulations as described above are now carried out for the second case, the measured texture of the beverage can aluminum alloy, see Tab. 7.4. At 50% thickness reduction and $\lambda^{\mathrm{HS}} = 0$, there are pronounced Cube as well as Brass components, but Cu and S are present, too. In the case of $\lambda^{\mathrm{HS}} = 0.99$, there are no cube components, and the peaks around the Brass component are scattered, while copper and S components are pronounced.
Consequently, as shown above, different textures can be obtained during the deformation process with an appropriate choice of λ^{HS}.

Table 7.3: Evolution of crystallographic texture for 50% thickness reduction (th. red.): rolling of quasi-isotropic initial texture (characteristic texture components: • Goss, ★ Cu, ▲ S, ■ Cube, ♦ Bs)

th.red.	λ^{HS}=0.99	λ^{HS}=0
0%	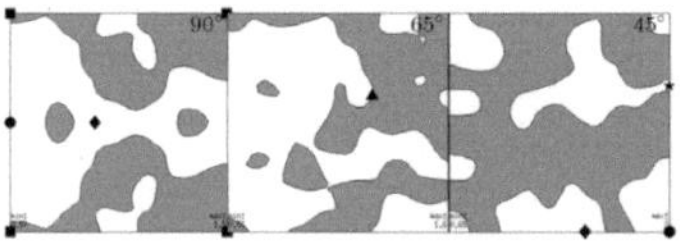	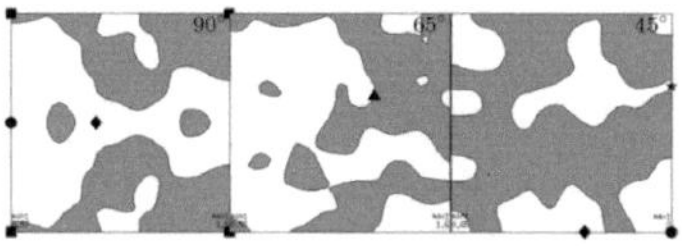
50%	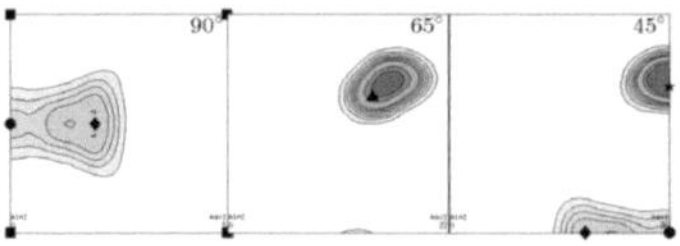	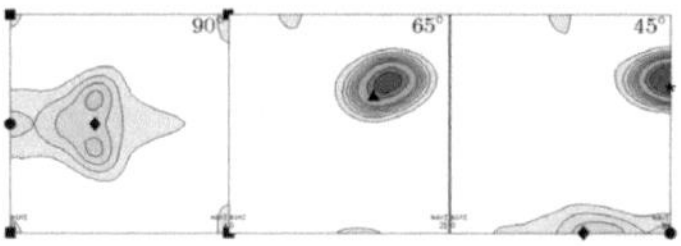

Table 7.4: Evolution of crystallographic texture for 50% thickness reduction (th. red.): rolling of EN AW-3104 initial texture (characteristic texture components: • Goss, ★ Cu, ▲ S, ■ Cube, ♦ Bs)

th.red.	λ^{HS}=0.99	λ^{HS}=0
0%	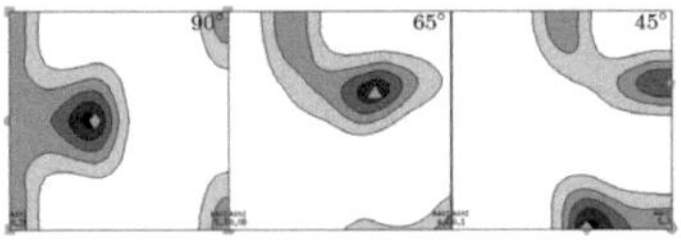	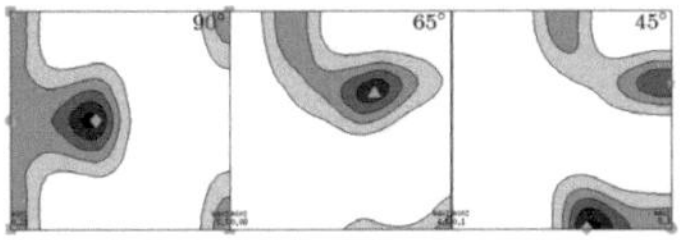
50%	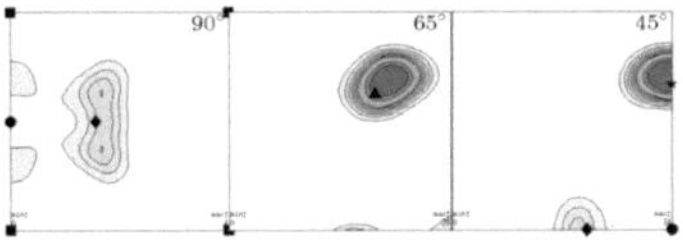	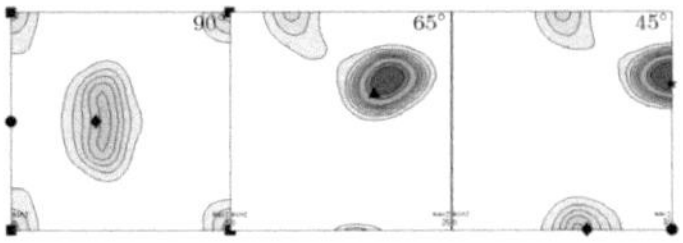

7.4 Characteristic orientations in deformation texture

In order to further investigate the influence of the homogenization on deformation texture, the aforementioned rolling process is focused on. A quasi-isotropic texture is subjected to plane strain compression. Characteristic orientations can be used to describe textures. Here, a certain set of them is used, in form of an orthogonal matrix:

$$\boldsymbol{Q}_\alpha, \; \alpha \in \{\mathrm{S}, \mathrm{Cu}, \mathrm{Cube}, \mathrm{Bs}, \mathrm{Goss}\} \;, \tag{7.2}$$

with the values for the orientations summarized in Tab. 3.1. For each of the calculated orientations in the simulation, $\boldsymbol{Q}_c^i$, $i = 1, \ldots, n_{\mathrm{ori}}$, the misorientation angle is computed by

$$\cos(\omega_{\mathrm{mis}}) = \frac{1}{2}\left(\mathrm{tr}(\Delta\boldsymbol{Q}) - 1\right) \;, \quad \text{with } \Delta\boldsymbol{Q} = \boldsymbol{Q}_\alpha {\boldsymbol{Q}_c^i}^{\mathsf{T}} \;. \tag{7.3}$$

Subsequently, for all symmetric transformations of one orientation $\boldsymbol{Q}_c^i$, the calculated minimal misorientation angle $\omega_{\mathrm{mis}}^{\mathrm{min}}$ is determined. In Fig. 7.4 –

7.8, all the misorientations for a $+/-5^\circ$ threshold are plotted over the subjected deformation.
For the Sachs type homogenization ($\lambda^{\mathrm{HS}} = 1.0$), there is a peak in volume fraction for the Goss component in Fig. 7.4. The Cube component in Fig. 7.5 completely vanishes for a Sachs type homogenization after 0.5 equivalent strain, the more Taylor like the homogenization is the higher is the volume fraction of cube components. In Fig. 7.6, the volume fraction of the S component for $\lambda^{\mathrm{HS}} = 1.0$ at equivalent strain equal to one has a maximum and then completely vanishes for 1.5 equivalent strain. The development for the Copper component in Fig. 7.7 is monotonously increasing for all homogenization parameters, for Sachs type deformation the formation of Copper components starts at 1.0 equivalent strain contrary to the almost immediate formation in the transition to Taylor type texture. The Brass component volume fraction in Fig. 7.8 has a distinct peak at 1.2 equivalent strain for $\lambda^{\mathrm{HS}} = 1.0$. For larger deformations applied, this component completely vanishes.

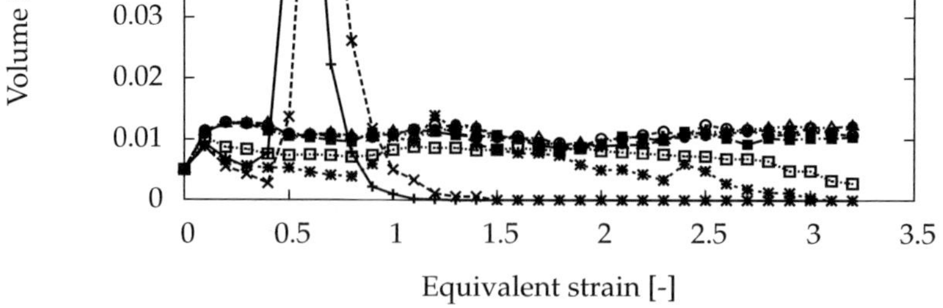

Figure 7.4: Goss component with misorientation angle $\omega_{\mathrm{mis}} = +/-5^\circ$ and varying homogenization parameter λ^{HS}

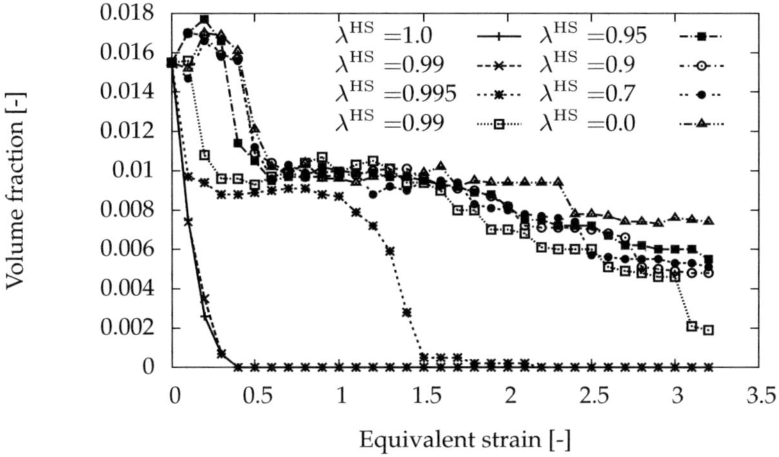

Figure 7.5: Cube component with misorientation angle $\omega_{\text{mis}} = +/-5°$ and varying homogenization parameter λ^{HS}

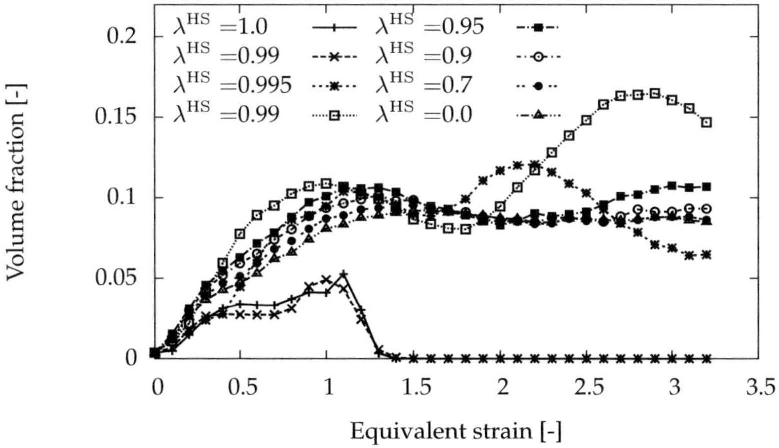

Figure 7.6: S component with misorientation angle $\omega_{\text{mis}} = +/-5°$ and varying homogenization parameter λ^{HS}

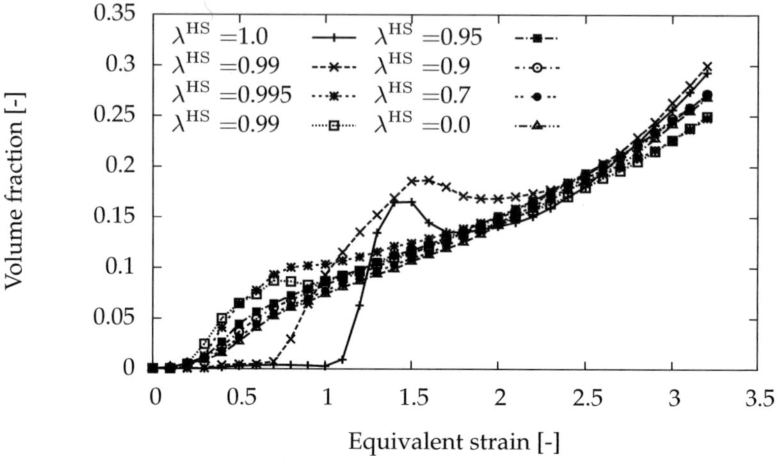

Figure 7.7: Cu component with misorientation angle $\omega_{\text{mis}} = +/-5°$ and varying homogenization parameter λ^{HS}

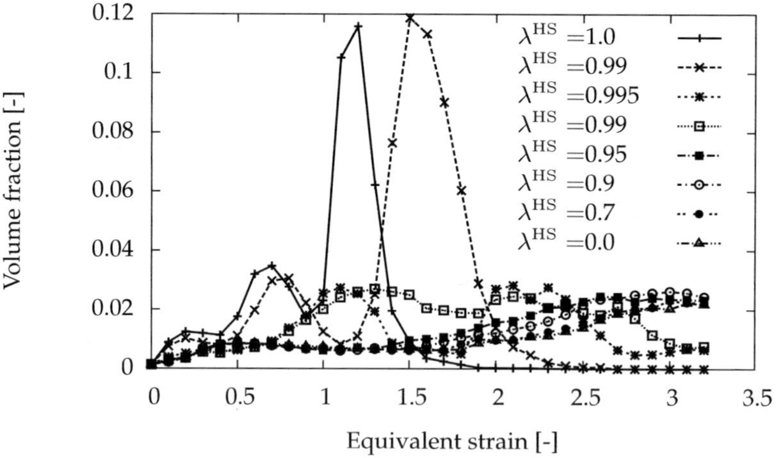

Figure 7.8: Brass component with misorientation angle $\omega_{\text{mis}} = +/-5°$ and varying homogenization parameter λ^{HS}

7.5 Sensitivity of homogenization parameter λ^{HS}

As shown in Appendix A, the homogenization parameter λ^{HS} is calculated with eq. (A.19) and its domain is $\lambda^{\mathrm{HS}} \in [0,1]$, with the first-order bounds by Voigt ($\lambda^{\mathrm{HS}} = 0$) and by Reuss ($\lambda^{\mathrm{HS}} = 1$). The question arises whether λ^{HS} is equally sensitive with respect to the reference stiffness $\mathbb{C}_0$ and its eigenvalue G_0 on the whole domain [0,1], or if there are certain ranges where a small deviation of G_0 has already a significant influence on the texture development.

By choosing a material, e.g., aluminum ($E = 70000$ MPa, $\nu = 0.3$), the shear modulus G and the bulk modulus K are known, as well. Due to the assumptions made in deriving the homogenization method, K_0 is also known and equal to K of aluminum. The only variable remaining in eq. (A.19) is the reference shear modulus G_0. Its influence on the homogenization parameter λ^{HS} is shown in Fig. 7.9. The gradient of λ^{HS} can be obtained by differentiating eq. (A.19) with respect to G_0, which leads to:

$$\frac{\mathrm{d}\lambda^{\mathrm{HS}}}{\mathrm{d}G_0} = -\frac{6G(16G_0^2 + 16G_0K_0 + 9K_0^2)}{(6G(2G_0 + K_0) + G_0(8G_0 + 9K_0))^2}\,, \tag{7.4}$$

with the limits

$$\lim_{G_0 \to 0} \frac{\mathrm{d}\lambda^{\mathrm{HS}}}{\mathrm{d}G_0} = -\frac{3}{2G}\,, \qquad \lim_{G_0 \to \infty} \frac{\mathrm{d}\lambda^{\mathrm{HS}}}{\mathrm{d}G_0} = 0\,. \tag{7.5}$$

The gradient of λ^{HS} is shown in Fig. 7.10. By jointly interpreting Fig. 7.9 and Fig. 7.10, it is concluded that λ^{HS} is, in fact, more sensitive to changes in G_0 in the vicinity of the Reuss bound ($\lambda^{\mathrm{HS}} = 1$), where small changes in λ^{HS} can already imply a substantial effect on the texture evolution.

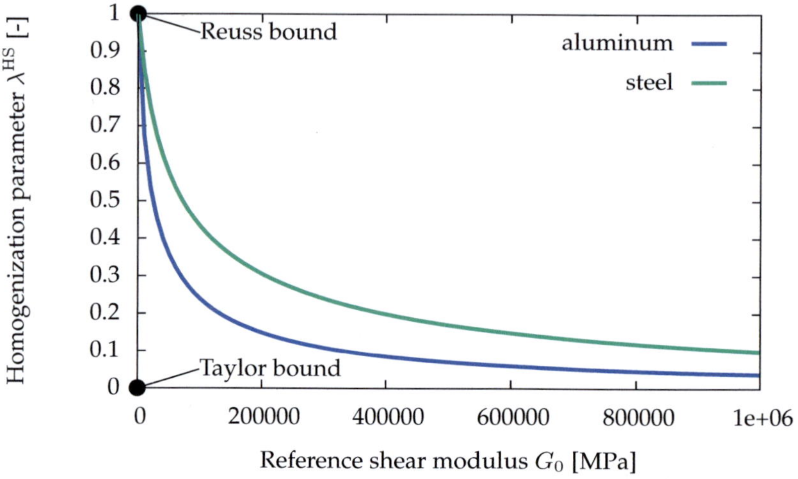

Figure 7.9: Influence of varying reference stiffness on λ^{HS}

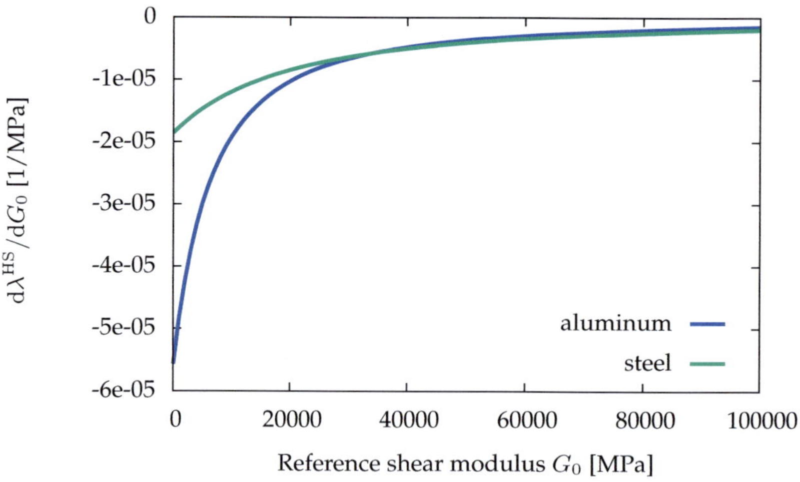

Figure 7.10: Influence of varying reference stiffness on $d\lambda^{HS}/dG_0$

Chapter 8

Multi-scale simulations based on crystal plasticity with non-linear Hashin-Shtrikman type homogenization

8.1 Forming simulations

8.1.1 Deep-drawing of BUP

The crystal plasticity constitutive equations described in Section 3.2 with the HS homogenization are applied to a deep-drawing of aluminum sheets in this section.

The BUP cup is a small-scale experiment of the beverage can deep-drawing process in order to investigate the earing behavior of the aluminum alloy. In the experimental setting, a circular blank ($d = 64$ mm, $t = 0.264$ mm) is set on top of a die and hold in place with a blank-holder (applied blank-holder force 8 kN). The punch displacement is applied until the sheet is completely drawn-in and a cup is formed. For the simulation, the same set up is used as for the above described experiment. Texture measurements of the aluminum alloy EN AW-3104 by Constellium are used as initial conditions for the crystallographic texture at the Gauss point level. To reduce the computation time, a reduced data set of EN AW-3104 is chosen with 48×4 orientations and the homogenization parameter $\lambda^{\mathrm{HS}} = 0$. The simulations are carried out with ABAQUS Explicit, with a mesh consisting of 1074 finite elements. The computational time (using identical time increments) compared to a ABAQUS von Mises plasticity model (and the same amount of CPUs) is about 150 times higher. This increase can be explained by the loop over 96 texture components at every Gauss point, where the material law must be evaluated for all 12 slip systems and with the rotation update performed by matrix-multiplication operations. The simulation is contrasted with the experimental earing profile in Fig. 8.1. There, the simulation predicts the number of ears and their location (with respect to the RD) accurately. Furthermore, the cup height, in general, is underestimated, and the earing behavior is overestimated significantly.

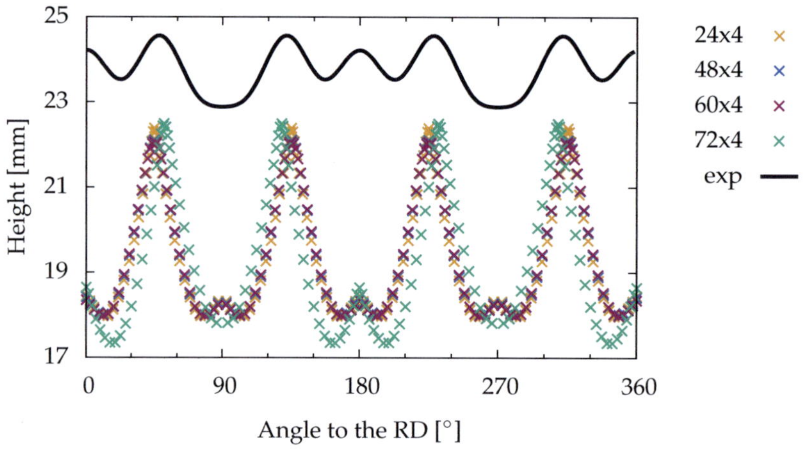

Figure 8.1: Influence of number of texture components on earing profile (with $\lambda^{\mathrm{HS}} = 0$): cup height of BUP vs angle to the RD

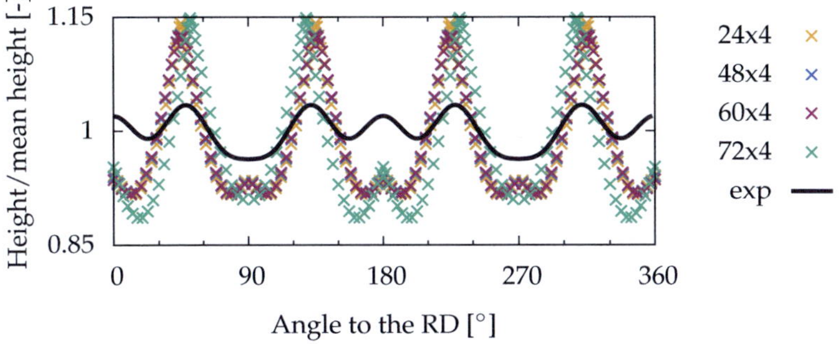

Figure 8.2: Influence of number of texture components on earing profile (with $\lambda^{\mathrm{HS}} = 0$): normalized cup height of BUP vs angle to the RD

In Fig. 8.2, where the cup height is normalized to the mean value of the cup height, this behavior is highlighted even more. In the following, a non Taylor homogenization parameter, i.e. $\lambda^{\mathrm{HS}} > 0$, and its influence on the earing profile is investigated. Figure 8.3 shows the influence of λ^{HS} on the earing profile after deep-drawing for a texture with 24×4 texture components. In a relatively large part of the homogenization parameter domain, for $0 \leq \lambda^{\mathrm{HS}} < 0.9$, the resulting earing profiles remain very similar to each other. With $\lambda^{\mathrm{HS}} \geq 0.99$, the absolute cup height decreases and the earing behavior, in general, is less pronounced than in the aforementioned cases. The results for $45°$ to the RD improve whereas the local maximum at $0°$ to the RD cannot be predicted anymore. Furthermore, the curvature of the earing profile for $45°$ to the RD with $1 - 10^{-4} \leq \lambda^{\mathrm{HS}} < 1 - 10^{-5}$ shows an opposing trend to the experimental results. Furthermore, the measured earing profile is shown, together with the Hill prediction as well as the best matching micro-mechanical prediction with $\lambda^{\mathrm{HS}} = 1 - 10^{-3}$, in Fig. 8.4. Therein, the percentage variance, in terms of area difference, between the simulations and the experiment is shown as well. The phenomenological Hill prediction has a $\approx 50\%$ higher area between the experiment compared to the micromechanical approach.

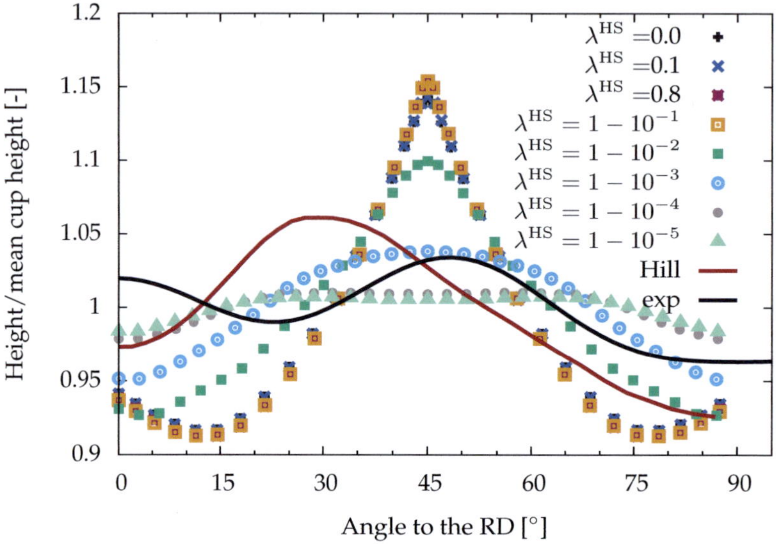

Figure 8.3: Earing profile: normalized cup height vs angle to RD (BUP)

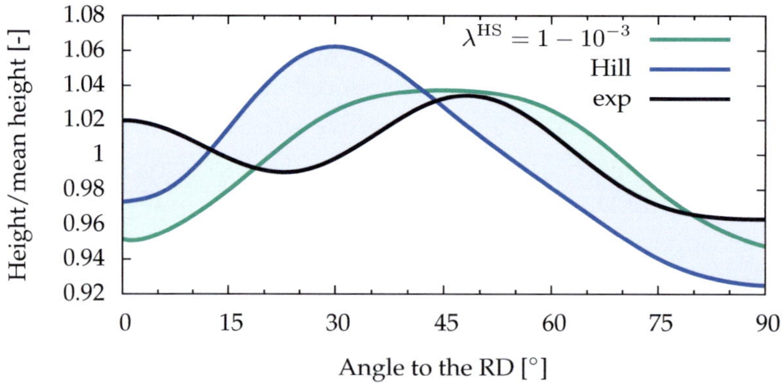

Figure 8.4: Earing profile BUP drawing: normalized cup height vs angle to the RD

8.1.2 Deep drawing of a cup

The beverage can cup drawing experiment is performed similarly to the aforementioned BUP cup drawing. Analogously, a circular blank ($d = 139.52$ mm, $t = 0.264$ mm) is set on top of a die and hold in place with a blank-holder (applied blank-holder force 15kN). The texture measurements of the aluminum alloy EN AW-3104 are used as initial conditions for the texture at the Gauss point level. To reduce the computational time, a reduced data set of EN AW-3104 is chosen with 24×4 orientations and tested with a varying homogenization parameter λ^{HS}. The simulations are carried out with ABAQUS Explicit, with a mesh consisting of 1145 finite elements. The results are shown in Fig. 8.5 and Fig. 8.6.

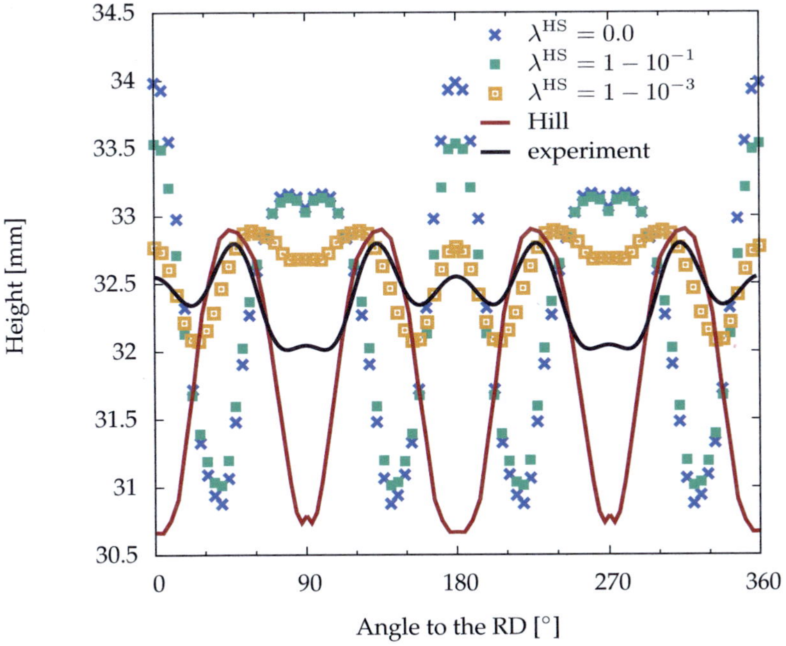

Figure 8.5: Earing profile CUP 1 drawing: cup height vs angle to the RD

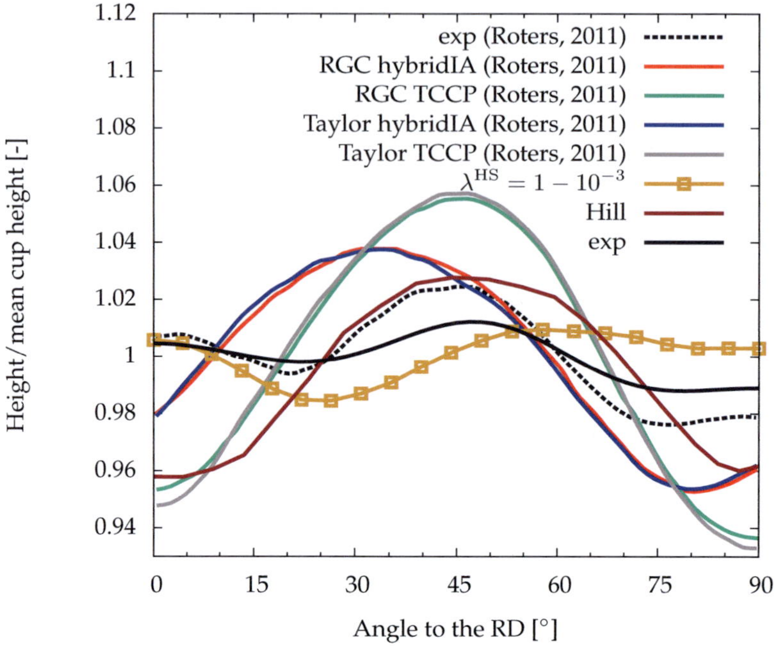

Figure 8.6: Earing profile CUP 1 drawing, comparison of simulations, experiments and results of Roters (2011): normalized cup height vs angle to the RD

Contrary to the BUP results, the mean cup height is approximated adequately in these simulations (with vanishing friction parameter) in Fig. 8.5. As in the aforementioned results, the simulations with $\lambda^{HS} = 0$ and $\lambda^{HS} = 1 - 10^{-1}$ overestimate the cup height in the RD significantly and do, as well, not predict the absolute cup height minimum at 90° to the RD. In the case of $\lambda^{HS} = 1 - 10^{-3}$, the cup height in the RD is approximated best by using this set of simulations. At 90° to the RD, the cup height shows a local minimum compared to the globally minimal cup height in the experimental results. In Fig. 8.6, the comparison of normalized cup height between the experimental earing profile and the phenomenological Hill as well as the best fit with $\lambda^{HS} = 1 - 10^{-3}$ of the micromechanical model is shown. These results are shown together with the results of Roters (2011), which was carried out for the same cold rolled aluminum alloy EN AW-3104. It should be noted that the experimental earing profile documented by Roters (2011) differs from the experiments in (Constellium CRV Test Report 43-11-1-3, 2013), as the maximum at 45° is higher and the minimum at 90° lower. However, the trend is captured best with the micromechanical model presented in this work with $\lambda^{\mathrm{HS}} = 1 - 10^{-3}$, capturing the behavior at 0° to the RD better than all other models.

8.1.3 Texture after deformation

The texture after deep-drawing in the RD and the TD is shown in Tab. 8.1. As previously shown in Fig. 6.14 and 6.15, the texture has been measured in the middle of the cup height. The simulation results were taken as mean value of the Gauss points of the corresponding element according to the cup height. The initially reduced texture of 24×4 orientations (cf. data reduction in Tab. 7.2) has been chosen mainly for numerical reasons in order to obtain simulation results in a managable amount of time.
In the top row of Tab. 8.1, the experimentally measured texture is shown. While in the RD the typical rolling texture, although with highest distribution density at the poles, is evident in the (111) pole figure. The texture in the TD is significantly different. On the left-hand side of the table, the results in the RD are shown. There, the CODF sections of the sample in the RD show a pronounced peak at the section with $\varphi_2 = 20°$ which cannot be predicted by either of the simulations. While there is a pronounced peak in the $\lambda^{\mathrm{HS}} = 1 - 10^{-3}$ simulation, it is, however, shifted by 30° into the negative Φ-direction. Furthermore, the section at $\varphi_2 = 45°$ has its maximal value at 90° and 30° in Φ- and φ_1-direction, respectively. There is a small peak present in the simulation with the largest homogenization parameter, whereas other components at Φ and φ_1 equal 90° are not observable in the simulations. The highest value of

the CODF in the simulations is at $\Phi = 0°$ and $\varphi_1 = 90°$. The high intensity peak of the experiment in this section is shifted to $\Phi = 30°$. In the section at $\varphi_2 = 65°$, the pronounced texture component is present in the Sachs type homogenization results. Moreover, the highest intesity in the simulations is at $\varphi_1 = 90$, but in the experiments this trend cannot be confirmed.
For the right-hand side of Tab. 8.1 the texture results in the TD are shown. Overall, the (111) pole figure is approximated acceptably by the simulations, with the best correspondance for $\lambda^{\mathrm{HS}} = 1 - 10^{-3}$. Compared to the undeformed texture, the brass component in the CODF section with $\varphi_2 = 20°$ has moved nearly $30°$ into the Φ-direction. This trend is captured with the homogenization parameter $\lambda^{\mathrm{HS}} = 1 - 10^{-3}$, although the orientation density is distributed between two peaks, the brass component peak and the stronger pronounced experimentally observed component peak. With a lower homogenization parameter value, the distribution is spread out more. The texture for Taylor type homogenization showing texture components nearly in cube component configuration. In the sections $\varphi_2 = 45°$ and $\varphi_2 = 65°$ there is a reasonably good approximation of the obtained texture component. Once again, the $\lambda^{\mathrm{HS}} = 1 - 10^{-3}$ homogenization parameter leads to more accurate results compared to the Taylor type solution of $\lambda^{\mathrm{HS}} = 0.0$.
The earing profile prediction, as well now as the texture prediction, confirm that the best suited value for the homogenization parameter is $\lambda^{\mathrm{HS}} = 1 - 10^{-3}$.

Table 8.1: Comparison of experiments with simulations of texture after deep-drawing at the middle of cup height in the RD and the TD: pole figures in (111)–direction as well as CODF sections of constant φ_2 (contours at $1, 2, 4, 6, 8, 10$ and 12)

	Texture at middle of cup height in the RD				Texture at middle of cup height in the TD			
	(111)	$\varphi_2 = 20°$	$\varphi_2 = 45°$	$\varphi_2 = 65°$	(111)	$\varphi_2 = 0°$	$\varphi_2 = 45°$	$\varphi_2 = 65°$
Experiment	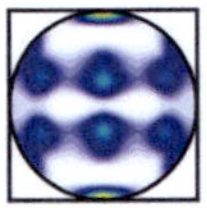	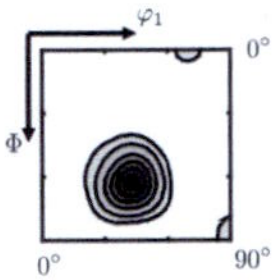	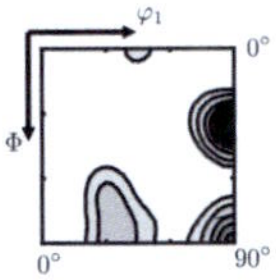	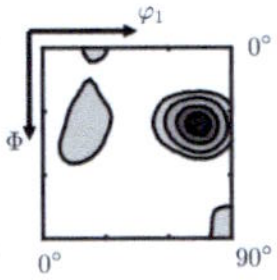	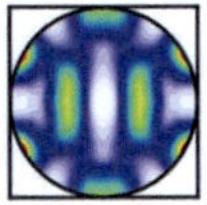	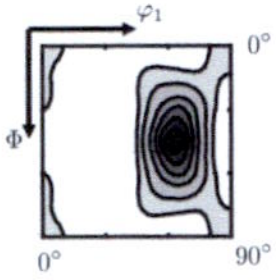	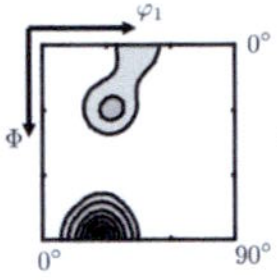	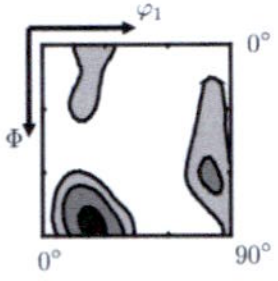
λ_{HS}=0.999	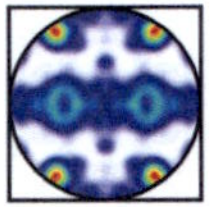	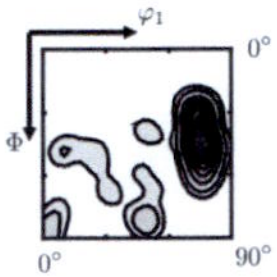	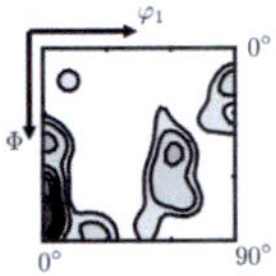	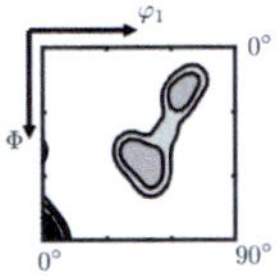	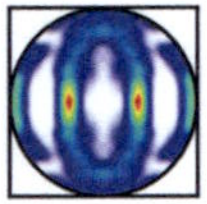	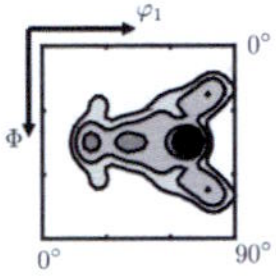	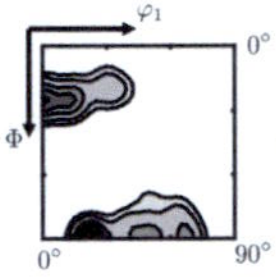	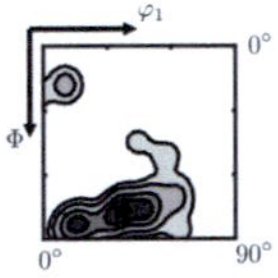
λ_{HS}=0.9	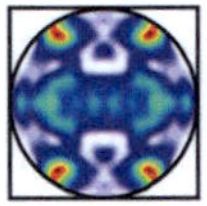	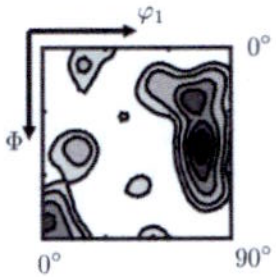	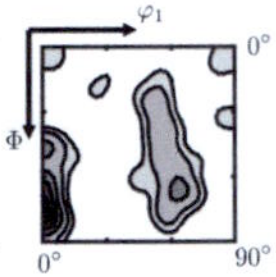	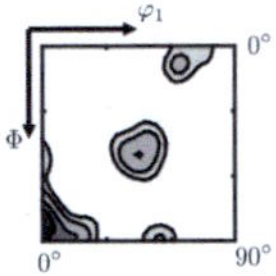	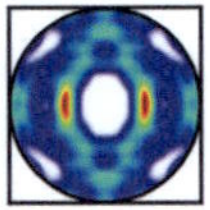	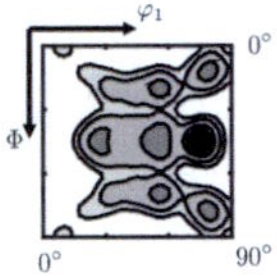	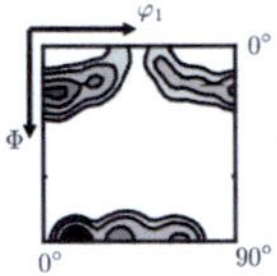	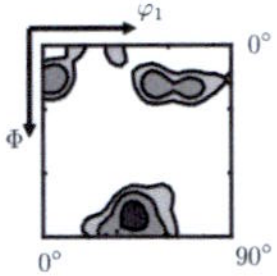
λ_{HS}=0	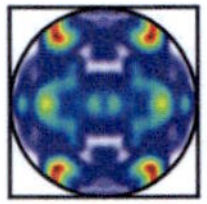	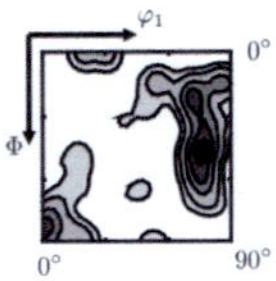	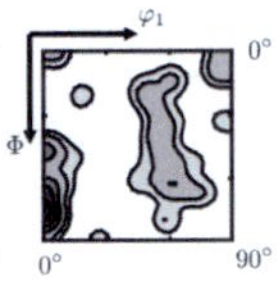	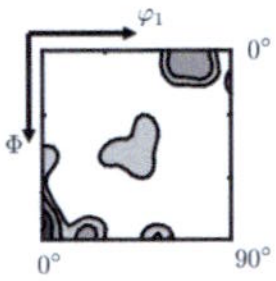	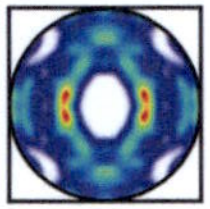	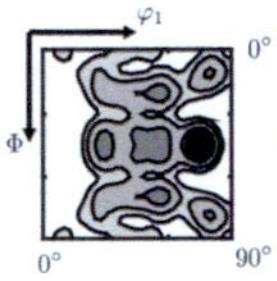	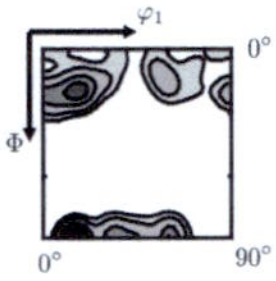	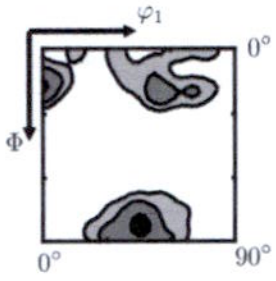

8.2 Localization

8.2.1 Tensile tests

From the tensile tests on sheet stripes (length $l = 300$ mm, width $w = 15$ mm and thickness $t = 0.264$ mm) described in Section 6.2 the fracture angles are shown in Fig. 8.7, 8.8, and 8.9 for the specimen orientations of $0°$, $45°$, and $90°$ to the RD, respectively.
The tensile test simulations are performed on models of sheet-metal stripes according to the experimental specimens. The polycrystalline model, with HS homogenization and 24×4 orientations, is used throughout the simulations. A heterogeneous mesh size is chosen to achieve a high resolution at locations where the localization is expected.

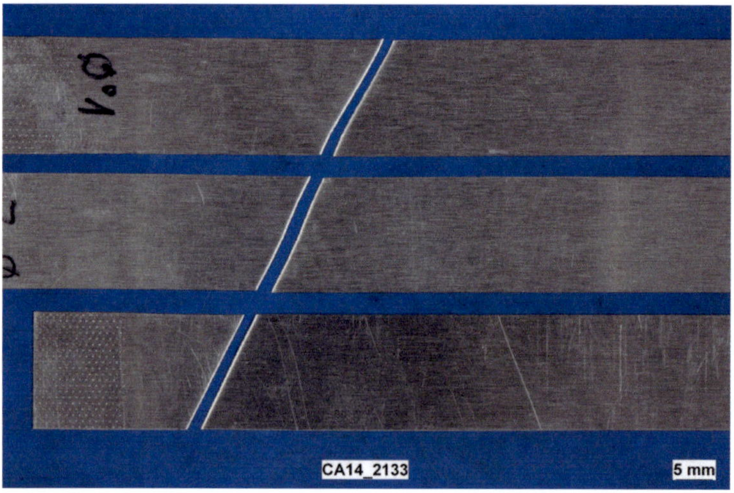

Figure 8.7: Three tested specimens in $0°$ to the RD (Suisse Technology Partners AG Test Report M-13-005, 2013)

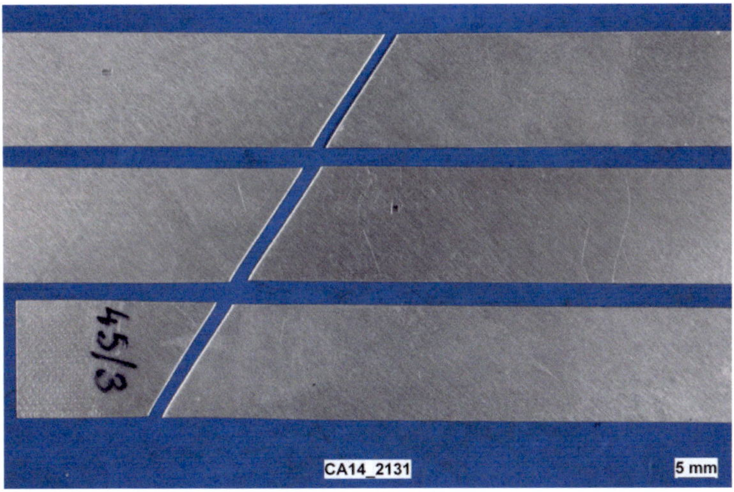

Figure 8.8: Three tested specimens in 45° to the RD (Suisse Technology Partners AG Test Report M-13-005, 2013)

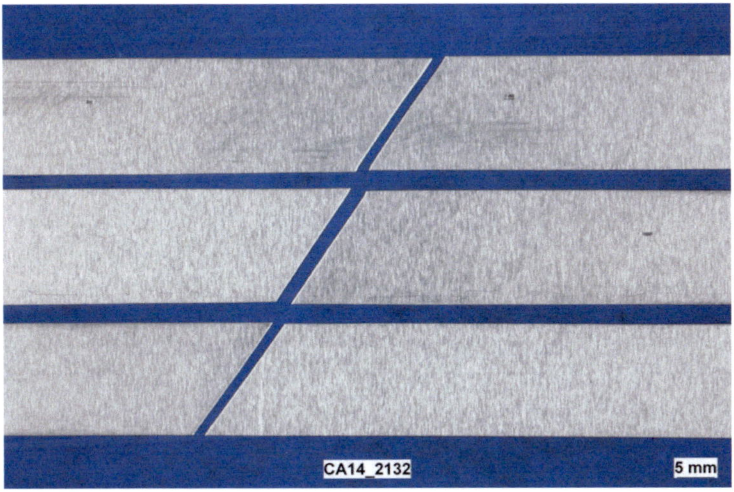

Figure 8.9: Three tested specimens in 90° to the RD (Suisse Technology Partners AG Test Report M-13-005, 2013)

The FE mesh can be seen in Fig. 8.10, where three elements used along the thickness-direction of the sheet. The localization is triggered by a decreased flow stress in a small set of elements in the geometric center of the specimen.

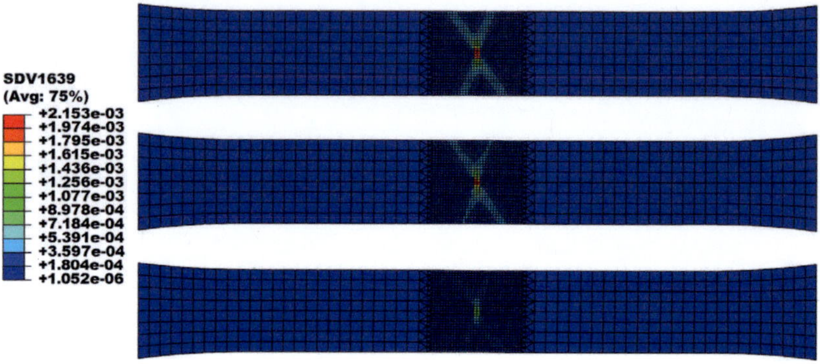

Figure 8.10: Localization behavior tensile test in 0° to the RD: equivalent plastic strain (parameter identification with $\lambda^{\mathrm{HS}}_{PI,I} = 0.0$) for $\lambda^{\mathrm{HS}} = 0.0, 0.5, 0.9$ (from top to bottom)

For the experimental data, three specimens were taken per each orientation (0°, 45° and 90° to the RD) of the sheet. The fracture angle is averaged over these three orientations and shown in Fig. 8.11 with the corresponding error bars. Localization angles for the example of 0° to the RD and $\lambda^{\mathrm{HS}}_{\mathrm{PI}} = 0$ are shown in Fig. 8.10. Other simulations for the different directions have been performed. These results for the localization angles, as well as the fracture angles from the experiments in Fig. 8.7, 8.8, and 8.9, are summarized in Fig. 8.11. Therein, at 0° to the RD, the localization angle is underestimated throughout the variation of homogenization parameters (deviation $5 - 25\%$) with the best fit obtained for $\lambda^{\mathrm{HS}}_{\mathrm{PI,I}} = 0$. At 45° to the RD, the best prediction is achieved with this parameter, too, in comparison to the experiments. The results for $\lambda^{\mathrm{HS}}_{\mathrm{PI,II}} = 0.999$ (Reuss type) as well as the $\lambda^{\mathrm{HS}} = 0$ solution of $\lambda^{\mathrm{HS}}_{\mathrm{PI,I}} = 0.0$ (Taylor type) is in good agreement with the experiment. Furthermore, the results of $\lambda^{\text{HS}} = 0.5$ and $\lambda^{\text{HS}} = 1 - 10^{-1}$ deviate by $\sim 12\%$ compared to the localization in the experiments.

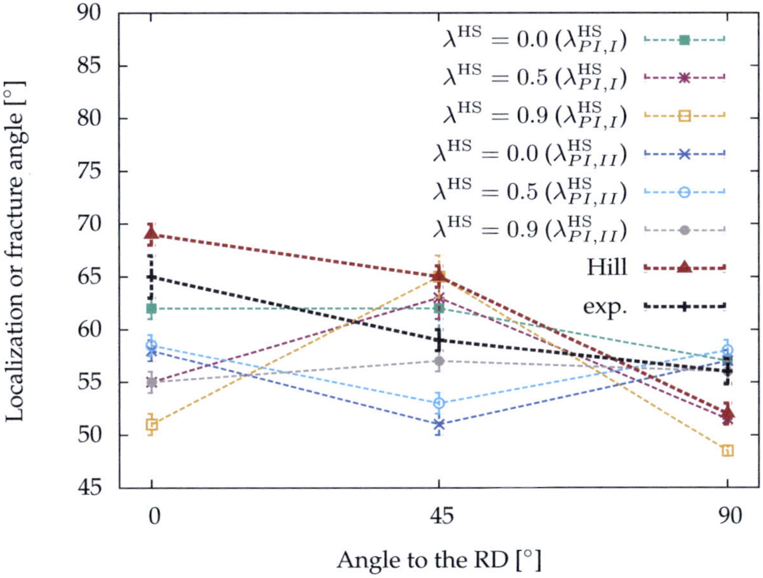

Figure 8.11: Localization angle: Experiments (Suisse Technology Partners AG Test Report M-13-005, 2013) and simulation for varying homogenization parameter λ^{HS} with different parameter identification (λ^{HS}_{PI})

In Fig. 8.12, the critical strain in terms of the localization of the FE solution is shown with respect to the parameter identification process, the homogenization parameter, and the angle to rolling direction. Generally, with an increasing homogenization parameter λ^{HS} from 0 to 1, the critical strain increases, with a fairly linear behavior between $\lambda^{\text{HS}} = 0$ and $\lambda^{\text{HS}} = 0.5$, and a non-linear behavior in the rest of the domain. The most sensitive behavior is again detected between $\lambda^{\text{HS}} = 1 - 10^{-1}$ and $\lambda^{\text{HS}} = 1.0$, as in the previous case, too. With the parameter identification $\lambda^{\text{HS}}_{\text{PI,I}}$, the results between $0°$, $45°$, and $90°$ to the RD are almost identical.

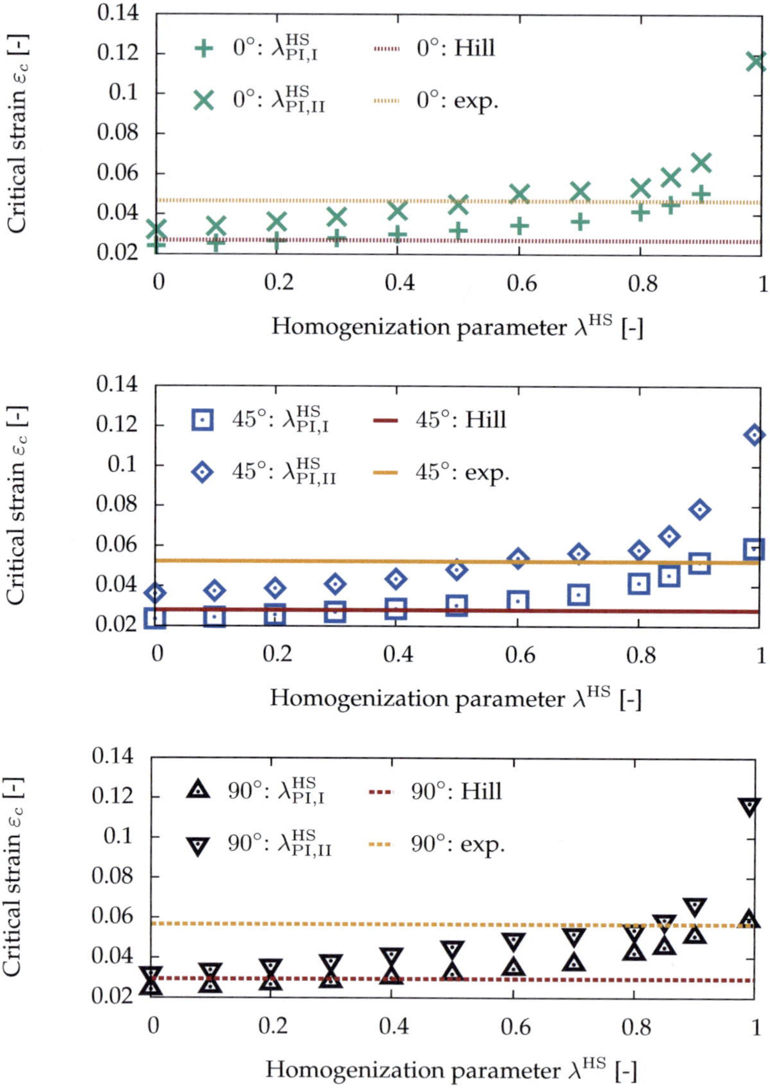

Figure 8.12: Critical strain over homogenization parameter for tension test simulation in 0° (top), 45° (middle), and 90° (bottom) to the RD: Comparison to Hill and experimental results (Suisse Technology Partners AG Test Report M-13-005, 2013)

Furthermore, with the parameter identification $\lambda^{\mathrm{HS}}_{\mathrm{PI,II}}$, the critical strains corresponding to the homogenization parameter have a generally larger regime than for the latter parameter identification procedure. In this case, the results in $45°$ to the RD differ from the experimental solutions.

The localization in the FE solution is evaluated for all different directions to the RD in Fig. 8.13, where the critical strain increases with increasing angle to the RD in the experiment. This trend is captured best with $\lambda^{\mathrm{HS}} = 0.99$, although the strain at $0°$ and $45°$ to the RD is overestimated.

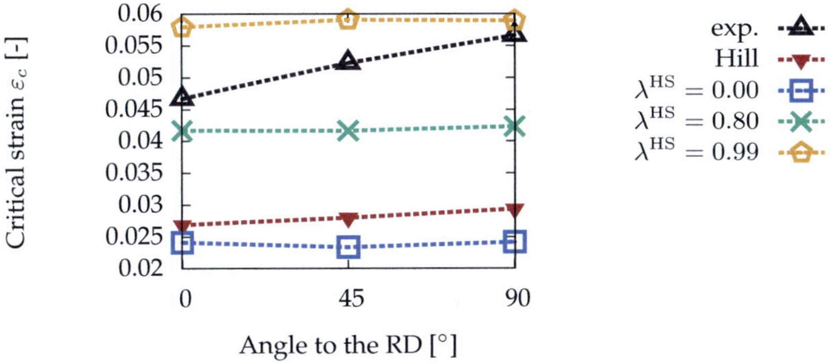

Figure 8.13: Critical strain over angle to RD: parameters identified with $\lambda^{\mathrm{HS}}_{\mathrm{PI,I}}$ and experiments according to Suisse Technology Partners AG Test Report M-13-005 (2013)

8.2.2 Nakajima tests

In order to determine the formability of the aluminum alloy a so called forming limit diagram (FLD) and forming limit curve (FLC) are used. These curves show the deformation localization for certain principle strain ratios. Nakajima tests are used to generate the FLDs, where the strains in the specimen plane (major and minor strain) are being evaluated.

All experiments shown here have been performed by C. Leppin (Suisse Technology Partners AG) on behalf of the industrial partner Constellium. In a Nakajima test, a fixed blank is being deformed until failure by a punch with a hemispherical top surface. Here, three geometries have been used (Geometry 1: $w = 40$ mm, Geometry 2: $w = 115$ mm and Geometry 3: $w = 168$ mm) with w describing the width of the specimen. Three specimens have been used for each geometry and the specimens after failure are shown in Tab. 8.2 a) – c) for Geometry 1, Tab. 8.2 d) – f) for Geometry 2 and Tab. 8.2 g) – h) for Geometry 3.
For the simulations, the localization is triggered by a (slightly) reduced flow stress in four elements in the center of the circular specimen. The simulations are carried out with the micromechanical model with Hashin-Shtrikman homogenization and are compared to the phenomenological solution using the Hill model. In Fig. 8.14, the experimental results of the Nakajima tests and the simulation results are both shown in one picture. The mean over four elements is shown for the values obtained by the simulation. More detailed FLDs are given in Fig. 8.15 – 8.17, where for each of the three geometries the experimental results as well as the localization initiation ($+$) and after localization ($\diamond$) are shown for different homogenization parameters.
For all of the three geometries, the Hill model underestimates the major strain compared to the experiment. For the micromechanical simulations, the localization initiation is under the FLC and after localization above the FLC. In conclusion, with a suitable choice of the homogenization parameter λ^{HS}, the experimental results can be predicted, whereas it is not possible to achieve a similarly good prediction with model of Hill.

Table 8.2: Experimental results Nakajima test with three specimens for each geometry: Geometry 1 a) – c), Geometry 2 d) – f) and Geometry 3 g) – i) (Suisse Technology Partners AG Test Report E-13-030, 2013)

a) 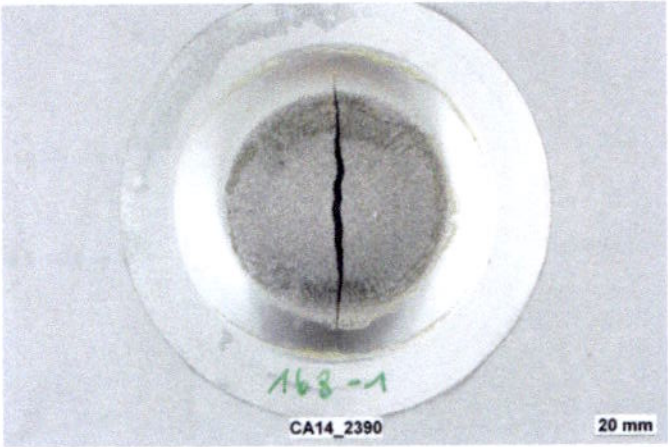

b) 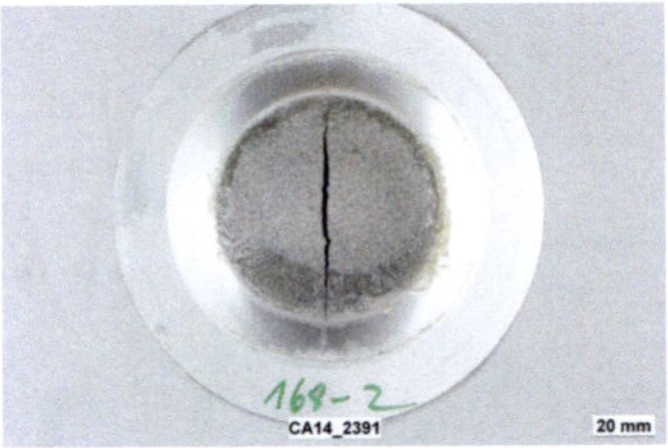

c) 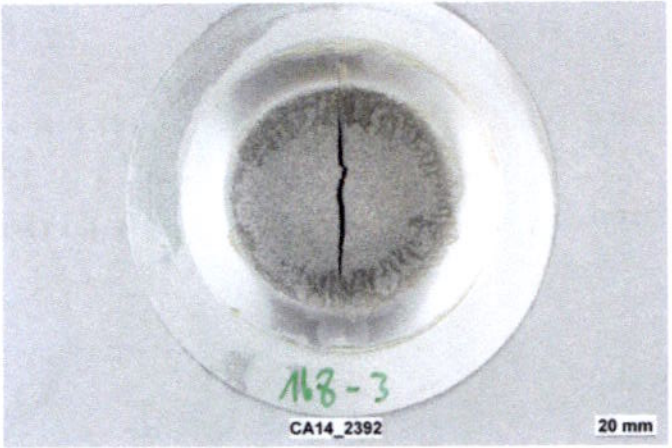

d) 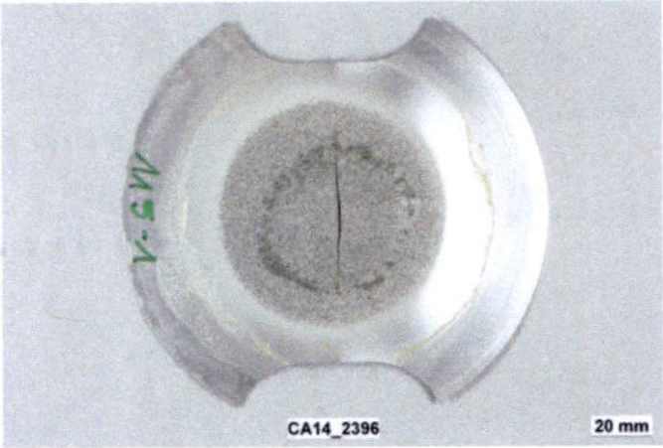

e) 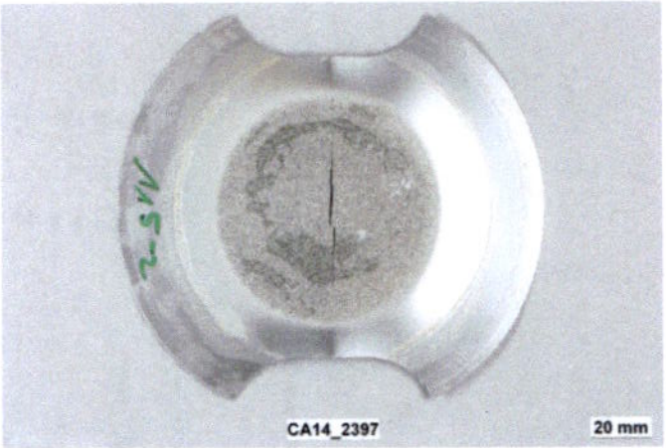

f) 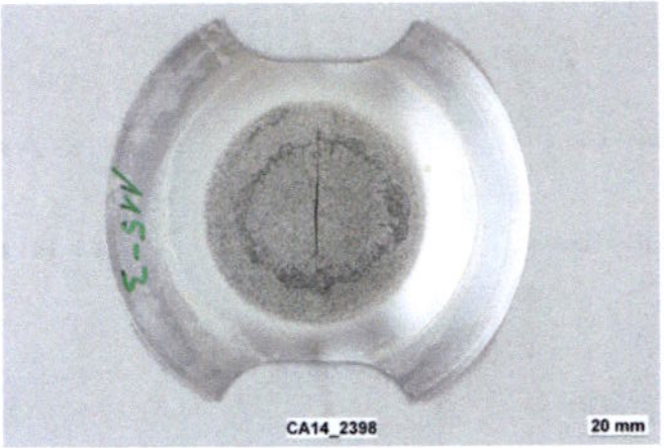

g) 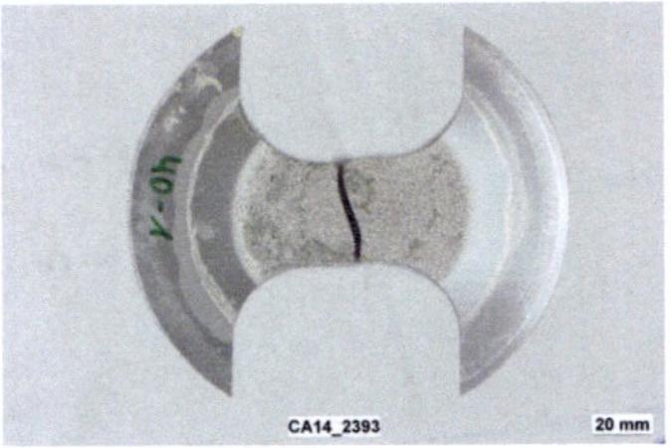

h) 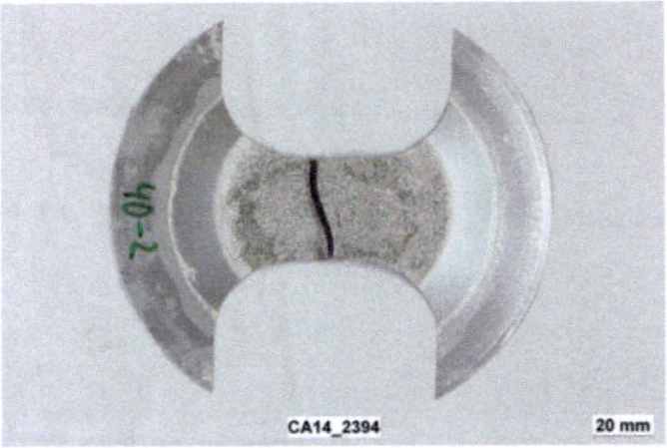

i) 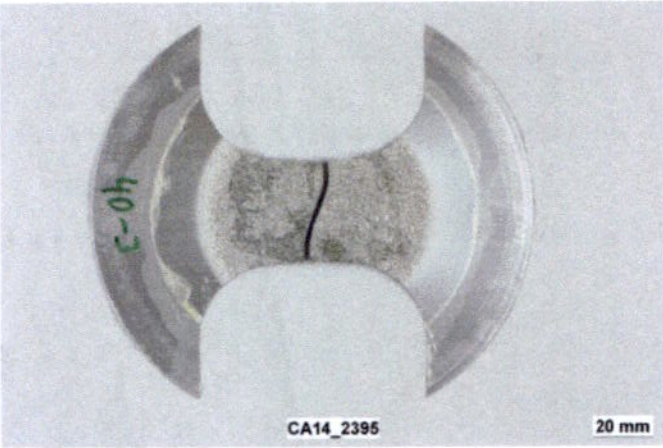

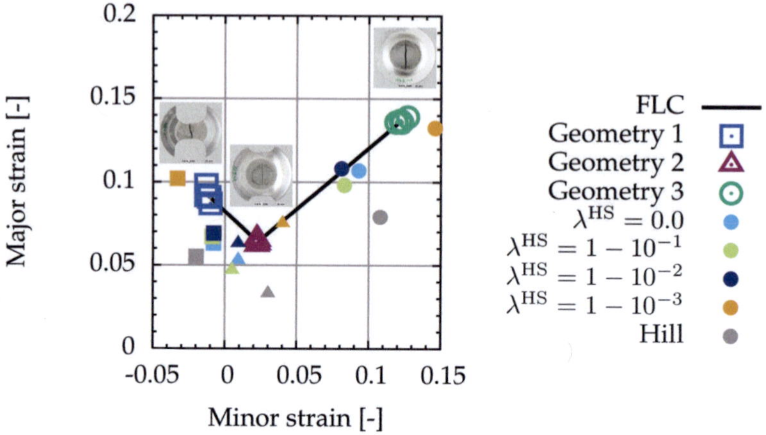

Figure 8.14: FLD - Experiments (Suisse Technology Partners AG Test Report E-13-030, 2013) and simulations for the three geometries in the Nakajima test

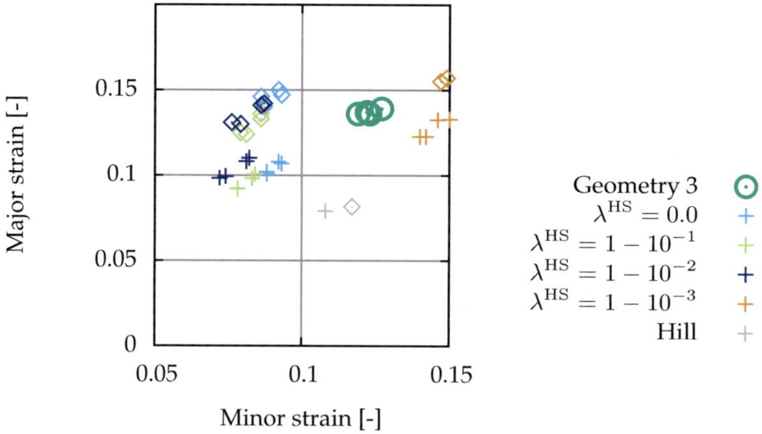

Figure 8.15: FLD - Experiments (Suisse Technology Partners AG Test Report E-13-030, 2013) and simulations (before (+) and after (◇) localization) for Geometry 3 in the Nakajima test

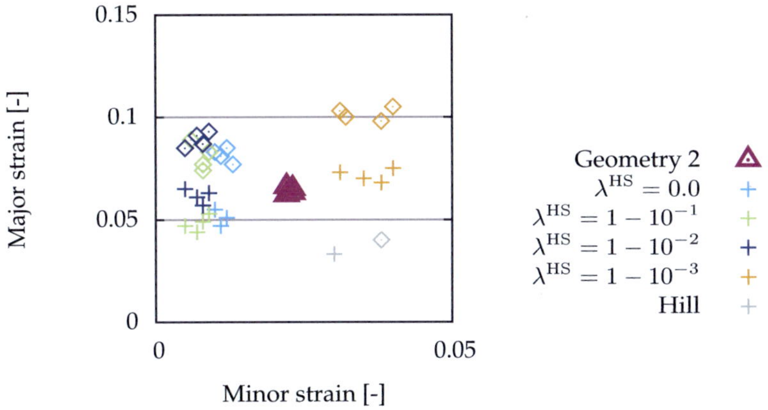

Figure 8.16: FLD - Experiments (Suisse Technology Partners AG Test Report E-13-030, 2013) and simulations (before (+) and after (◇) for Geometry 2 in the Nakajima test

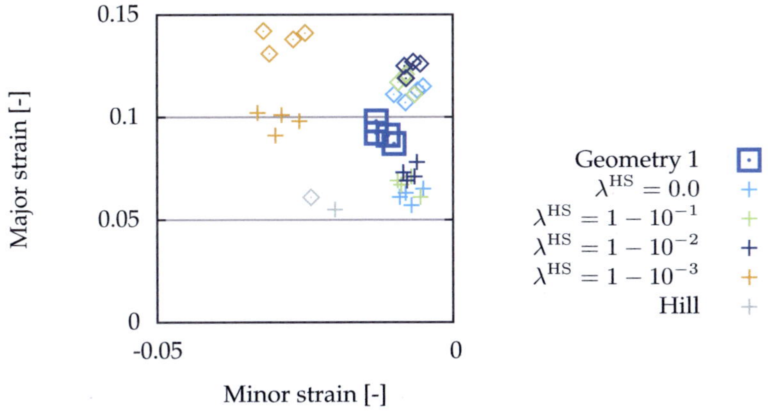

Figure 8.17: FLD - Experiments (Suisse Technology Partners AG Test Report E-13-030, 2013) and simulations (before (+) and after (◇) for Geometry 1 in the Nakajima test

Chapter 9

Summary

For forming simulations in industrial applications, usually phenomenological models are chosen to predict the texture development. This implies, however, that the microstructure development and the resulting plastic anisotropy cannot be simulated with these kind of models anymore. In order to take into account the microstructure evolution, a micro-mechanical approach for the constitutive modeling of sheet metal behavior during deep drawing has been formulated for aluminum (EN AW-3104). This particular material is often used in the packaging industry. The material model uses crystal plasticity theory and a non-linear homogenization scheme. In order to perform industrially applicable simulations, the constitutive model has been implemented in a commercial, explicit finite element tool (ABAQUS Explicit).

The application to a beverage can deep drawing process was chosen based on the fact that phenomenological models, there, produce unsatisfactory results. Furthermore, due to the high quantities in beverage can productions the reduction of scrap is also of great importance in this context. The deep drawing in the manufacturing process consists of two deep drawing steps followed by three ironing steps. In order to understand the material behavior, different kind of experiments (uniaxial tension tests in different directions to the rolling direction, texture measurements, evaluation of earing profiles, hydraulic bulge and Nakajima test) have been performed by the industrial partner Constellium. The hardening behavior has been identified with the hydraulic bulge tests.

Texture measurements have been used to specify the initial texture in the micromechanical model.
Due to the fact that the experimental orientation data, obtained with X-ray diffraction or EBSD measurements, is very large and the constitutive model is implemented at the Gauss point level, a reduction of this dataset is necessary. In this reduction method, all orientation data is transformed into fundamental zones in the Cartesian Euler angle space. This space is tessellated and averaged within the metrics of the Euler space. As a result, crystallographic texture of the material can be described by considerably less orientations. An important step is choosing a sufficient degree of compression, for the data set to be large enough for still being representative, and small enough to be used efficiently in an explicit finite element implementation.
Due to the nature of the aluminum alloy and the necessity for low computational costs, the homogenization scheme has been simplified for the case of elastically isotropic and plastically incompressible materials. Thereby, the distribution of stress and strain fluctuations is governed by a scalar parameter. This parameter varies between zero and one. The minimal and maximal values of the parameter give the classical simple bounds of Taylor and Sachs, respectively. Furthermore, the investigation showed that this homogenization parameter has a crucial influence in a rolling process simulation on the texture formation and on the stress level. Due to the fact that the parameter is deduced from an approximation of the stress field through constant stress polarizations directly from Green's function and is not introduced ad hoc, this is a new and promising way of considering stress and strain fluctuations. This novel mean field approach can be applied to other materials and other deformation processes as well.
For a deep drawing process of a circular blank, the micromechanical model was applied. As in the case of the rolling process, the homogenization parameter has a great impact on the non-uniform cup height (earing). The simulation results for Sachs type behavior predict the correct number of six ears, whereas the phenomenological Hill model leads to

inferior overall results and only predicts four ears. More sophisticated phenomenological models of the Barlat type (e.g., Aretz and Barlat, 2013) can predict the earing behavior but neither are they implemented in commercially used codes (e.g., Autoform) nor is their application straight forward due to the very high number (27) of material parameters. Therefore, these models are restricted to academia research and not used in industrial simulations. Another benefit of the micromechanical model is that the texture data is available after the cup drawing process and can be compared to experimentally obtained results. In comparison with the experiment, the simulation can predict the leading texture components, where the Sachs type behavior is again preferable to the Taylor type homogenization procedure. This result is particularly in good agreement with the experimentally measured pole figures and CODFs in transversal direction. Furthermore, the predictions of the new micromechanical model for the localization observed in the Nakajima test are preferable to the prediction by the commercially used Hill model.

Summarizing, this work presents a framework for incorporating information from microstructure to simulations for metal forming applications. A theory has been formulated for the special case of elastically isotropic media. This approach, together with data reduction and an efficient implementation in the framework of an explicit finite element scheme, offers the possibility to investigate behavior of a crystalline material (aluminum) under different boundary conditions. This two-scale approach offers new possibilities for sheet metal forming simulations, where conventional phenomenological models are no longer able to meet the demands of accurate predictions in industrial applications.

Appendix A

Derivation of the new effective medium equations for the polycrystal

As introduced in Section 5.2, an interaction law is given in eq. (5.9) for the non-linear Hashin-Shtrikman type homogenization scheme. In order to increase the readability, the subscript α is omitted in the following derivation. The Hashin-Shtrikman interaction law reads

$$\Delta\boldsymbol{\sigma} - \Delta\bar{\boldsymbol{\sigma}} = \mathbb{L}[\Delta\boldsymbol{\varepsilon} - \Delta\bar{\boldsymbol{\varepsilon}}] \, . \tag{A.1}$$

where the stress increment can be substituted by the incremental form of Hooke's law

$$\Delta\boldsymbol{\sigma} = \mathbb{C}\left[\Delta\boldsymbol{\varepsilon} - \Delta\boldsymbol{\varepsilon}_{\mathrm{p}}\right] \, . \tag{A.2}$$

For solving eq. (A.1), the following ansatz is taken from Jöchen (2013) for the strain localization

$$\Delta\boldsymbol{\varepsilon} = \mathbb{A}_{\mathrm{e}}\left[\Delta\bar{\boldsymbol{\varepsilon}}\right] - \boldsymbol{A}_{\mathrm{p}} \, . \tag{A.3}$$

Therein, the elastic strain localization tensor and the plastic strain localization tensor read

$$\mathbb{A}_{\mathrm{e}} = \mathbb{R}\langle\mathbb{R}\rangle^{-1} \, , \quad \boldsymbol{A}_{\mathrm{p}} = -\mathbb{R}\mathbb{C}\left[\Delta\boldsymbol{\varepsilon}_{\mathrm{p}}\right] + \mathbb{A}_{\mathrm{e}}\langle\mathbb{R}\mathbb{C}\left[\Delta\boldsymbol{\varepsilon}_{\mathrm{p}}\right]\rangle \, , \tag{A.4}$$

respectively. The fourth-order tensor $\mathbb{R}$, used in the strain localization tensors, is defined by

$$\mathbb{R} = \left(\mathbb{P}_0{}^{-1} + \delta\mathbb{C}\right)^{-1}, \tag{A.5}$$

with $\delta\mathbb{C} = \mathbb{C} - \mathbb{C}_0$. These equations hold under the assumption of constant stress polarization for a Hashin-Shtrikman localization.

In order to determine the governing equations, an elastically isotropic material behavior is assumed (for both $\mathbb{C}$ and $\mathbb{C}_0$). Furthermore, plastic incompressibility is assumed (i.e., $\mathbb{C} - \mathbb{C}_0 \sim \mathbb{P}_2^{\text{iso}}$). The fourth-order tensors $\mathbb{C}_0$ and $\mathbb{C}$ are then given by

$$\mathbb{C}_0 = c_1^0 \mathbb{P}_1^{\text{iso}} + c_2^0 \mathbb{P}_2^{\text{iso}}, \tag{A.6}$$

$$\mathbb{C} = c_1 \mathbb{P}_1^{\text{iso}} + c_2 \mathbb{P}_2^{\text{iso}}, \quad \text{with } c_1 = c_1^0 . \tag{A.7}$$

Hill's polarization tensor for the assumption of a spherical inclusion in an isotropic material reads (Mura, 1987; Gross and Seelig, 2011)

$$\mathbb{P}_0 = p_1 \mathbb{P}_1^{\text{iso}} + p_2 \mathbb{P}_2^{\text{iso}} \tag{A.8}$$

with the two factors

$$p_1 = \frac{1}{3K_0 + 4G_0} \quad \text{and} \quad p_2 = \frac{3(K_0 + 2G_0)}{5G_0(3K_0 + 4G_0)} . \tag{A.9}$$

With these assumptions it is implied that $\mathbb{R}$ is constant and homogeneous

$$\mathbb{R} = \left(\mathbb{P}_0{}^{-1} + (c_2 - c_2^0)\mathbb{P}_2^{\text{iso}}\right)^{-1}, \tag{A.10}$$

$$\mathbb{R} = m_1 \mathbb{P}_1^{\text{iso}} + m_2 \mathbb{P}_2^{\text{iso}}, \tag{A.11}$$

with the two factors

$$m_1 = p_1 \quad \text{and} \quad m_2 = \frac{1}{1/p_2 + c_2 - c_2^0} . \tag{A.12}$$

Consequently, $\langle\mathbb{R}\rangle = \mathbb{R}$ holds and the elastic strain localization tensor from eq. (A.4) reduces to

$$\mathbb{A}_{\mathrm{e}} = \mathbb{R}\langle\mathbb{R}\rangle^{-1} = \mathbb{R}\mathbb{R}^{-1} = \mathbb{I}^S \ . \tag{A.13}$$

The fourth-order composition of $\mathbb{R}$ and $\mathbb{C}$ is simplified with the equations (A.11) and (A.7), respectively, to

$$\mathbb{R}\mathbb{C} = m_1 c_1^0 \mathbb{P}_1^{\mathrm{iso}} + m_2 c_2 \mathbb{P}_2^{\mathrm{iso}} \ . \tag{A.14}$$

To obtain the plastic localization tensor $\boldsymbol{A}_{\mathrm{p}}$ from eq. (A.4), the results from equations (A.13) and (A.14) are taken into account as well as the deviatoric nature of the plastic strain increment ($\Delta\boldsymbol{\varepsilon}_{\mathrm{p}} = \Delta\boldsymbol{\varepsilon}_{\mathrm{p}}'$), leading to

$$\boldsymbol{A}_{\mathrm{p}} = m_2 c_2 \left(\Delta\boldsymbol{\varepsilon}_{\mathrm{p}} - \langle\Delta\boldsymbol{\varepsilon}_{\mathrm{p}}\rangle\right) \ . \tag{A.15}$$

Combining the results from eq. (A.13) and (A.15) gives the strain localization increment in eq. (A.3)

$$\Delta\boldsymbol{\varepsilon} = \mathbb{A}_{\mathrm{e}}\left[\Delta\bar{\boldsymbol{\varepsilon}}\right] - \boldsymbol{A}_{\mathrm{p}} = \Delta\bar{\boldsymbol{\varepsilon}} - m_2 c_2 \left(\Delta\boldsymbol{\varepsilon}_{\mathrm{p}} - \langle\Delta\boldsymbol{\varepsilon}_{\mathrm{p}}\rangle\right) \ . \tag{A.16}$$

Together with Hooke's law from eq. (A.2), the spherical and deviatoric stress increments are

$$\Delta\boldsymbol{\sigma}^0 = 3K\Delta\bar{\boldsymbol{\varepsilon}}^0 \ , \tag{A.17}$$

$$\Delta\boldsymbol{\sigma}' = 2G\big(\Delta\bar{\boldsymbol{\varepsilon}}' - \Delta\boldsymbol{\varepsilon}_{\mathrm{p}} + \underbrace{m_2 c_2}_{=:\lambda^{\mathrm{HS}}} \left(\Delta\boldsymbol{\varepsilon}_{\mathrm{p}} - \langle\Delta\boldsymbol{\varepsilon}_{\mathrm{p}}\rangle\right)\big) \ , \tag{A.18}$$

respectively, with the scalar homogenization parameter being defined as

$$\lambda^{\mathrm{HS}} = m_2 c_2 = \frac{6G(2G_0 + K_0)}{6G(2G_0 + K_0) + G_0(8G_0 + 9K_0)} \ . \tag{A.19}$$

It should be noted that under the assumptions made, a single scalar parameter λ^{HS} governs the localization relation. The domain of λ^{HS} is derived by considering the special cases for $G_0 \to 0$ and $G_0 \to \infty$:

$$\lim_{G_0 \to 0} \lambda^{\mathrm{HS}} = 1 \,, \tag{A.20}$$

$$\lim_{G_0 \to \infty} \lambda^{\mathrm{HS}} = 0 \,. \tag{A.21}$$

As shown in Fig. 7.9 in Section 7.5, the function of λ^{HS} has a global maximum (equal to one) for $G_0 \to 0$, then monotonically decreases to the global minimum (equal to zero) for $G_0 \to \infty$. Therefore, the domain for the homogenization parameter is $\lambda^{\mathrm{HS}} \in [0, 1]$. The simple bounds by Voigt and Reuss (Ponte Castañeda and Suquet, 1998) can be found for this Hashin-Shtrikman non-linear homogenization as follows:
Voigt ($\lambda^{\mathrm{HS}} = 0$)

$$\Delta\boldsymbol{\sigma}' = 2G\big(\Delta\bar{\boldsymbol{\varepsilon}}' - \Delta\boldsymbol{\varepsilon}_{\mathrm{p}}\big) \,, \tag{A.22}$$

Reuss ($\lambda^{\mathrm{HS}} = 1$)

$$\Delta\boldsymbol{\sigma}' = 2G\big(\Delta\bar{\boldsymbol{\varepsilon}}' - \langle\Delta\boldsymbol{\varepsilon}_{\mathrm{p}}\rangle\big) = \langle\Delta\boldsymbol{\sigma}'\rangle \,. \tag{A.23}$$

In eq. (A.22) the Voigt approximation with homogeneous strains is shown. Homogeneous stresses, as calculated by the volume average of the deviatoric stresses, lead to a Reuss type solution in eq. (A.23). Intermediate values of lambda give estimates between the Voigt and the Reuss estimate. The Hashin-Shtrikman type homogenization scheme $\big(\lambda^{\mathrm{HS}} \in (0, 1)\big)$ can be written as

$$\Delta\boldsymbol{\sigma}' = 2G\Big(\Delta\bar{\boldsymbol{\varepsilon}}' - \big((1 - \lambda^{\mathrm{HS}})\Delta\boldsymbol{\varepsilon}_{\mathrm{p}} + \lambda^{\mathrm{HS}}\langle\Delta\boldsymbol{\varepsilon}_{\mathrm{p}}\rangle\big)\Big) \,. \tag{A.24}$$

Appendix B

Anisotropic phenomenological material model: Hill yield criterion

One of the widely used yield criteria for anisotropic materials (for orthotropic symmetry with three orthogonal symmetry planes) was proposed by Hill (1948) (see also von Mises (1928)). A more detailed overview of the properties of the model described here is given in Banabic et al. (2000). The Hill yield surface is given by:

$$f(\boldsymbol{\sigma}) = \sqrt{\Sigma_{\mathrm{n}} + \Sigma_{\mathrm{s}}}\ , \tag{B.1}$$

$$\Sigma_{\mathrm{n}} = F\left(\sigma_{22} - \sigma_{33}\right)^2 + G\left(\sigma_{33} - \sigma_{11}\right)^2 + H\left(\sigma_{11} - \sigma_{22}\right)^2\ , \tag{B.2}$$

$$\Sigma_{\mathrm{s}} = 2\left(L\sigma_{23}^2 + M\sigma_{13}^2 + N\sigma_{12}^2\right)\ . \tag{B.3}$$

The material constants F, G, H, L, M and N are defined as:

$$F = \frac{\sigma_0^2}{2}\left(\frac{1}{\hat{\sigma}_{22}^2} + \frac{1}{\hat{\sigma}_{33}^2} - \frac{1}{\hat{\sigma}_{11}^2}\right) = \frac{1}{2}\left(\frac{1}{R_{22}^2} + \frac{1}{R_{33}^2} - \frac{1}{R_{11}^2}\right)\ , \tag{B.4}$$

$$G = \frac{\sigma_0^2}{2}\left(\frac{1}{\hat{\sigma}_{33}^2} + \frac{1}{\hat{\sigma}_{11}^2} - \frac{1}{\hat{\sigma}_{22}^2}\right) = \frac{1}{2}\left(\frac{1}{R_{33}^2} + \frac{1}{R_{11}^2} - \frac{1}{R_{22}^2}\right)\ , \tag{B.5}$$

$$H = \frac{\sigma_0^2}{2}\left(\frac{1}{\hat{\sigma}_{11}^2} + \frac{1}{\hat{\sigma}_{22}^2} - \frac{1}{\hat{\sigma}_{33}^2}\right) = \frac{1}{2}\left(\frac{1}{R_{11}^2} + \frac{1}{R_{22}^2} - \frac{1}{R_{33}^2}\right)\ , \tag{B.6}$$

$$L = \frac{3}{2}\left(\frac{\tau_0}{\hat{\sigma}_{23}}\right)^2 = \frac{3}{2}\frac{1}{R_{23}^2}\,, \tag{B.7}$$

$$M = \frac{3}{2}\left(\frac{\tau_0}{\hat{\sigma}_{13}}\right)^2 = \frac{3}{2}\frac{1}{R_{13}^2}\,, \tag{B.8}$$

$$N = \frac{3}{2}\left(\frac{\tau_0}{\hat{\sigma}_{12}}\right)^2 = \frac{3}{2}\frac{1}{R_{12}^2}\,, \tag{B.9}$$

where $\hat{\sigma}_{ij}$ is the measured yield stress when σ_{ij} is applied as the stress-component not equal to zero. Furthermore, σ_0 is a fixed but arbitrary reference yield stress, R_{ij} are anisotropic yield stress ratios and $\tau_0 = \sigma_0/\sqrt{3}$. The six yield stress ratios are defined as follows:

$$R_{11} = \frac{\hat{\sigma}_{11}}{\sigma_0}\,, \quad R_{22} = \frac{\hat{\sigma}_{22}}{\sigma_0}\,, \quad R_{33} = \frac{\hat{\sigma}_{33}}{\sigma_0}\,, \tag{B.10}$$

$$R_{12} = \frac{\hat{\sigma}_{12}}{\tau_0}\,, \quad R_{13} = \frac{\hat{\sigma}_{13}}{\tau_0}\,, \quad R_{23} = \frac{\hat{\sigma}_{23}}{\tau_0}\,. \tag{B.11}$$

They must be specified for the use of the material model in Abaqus (Simulia, 2012).

Hill's yield surface is defined through parameters R_{ij} for the stress ratios describing the onset of plasticity. However, in sheet metal forming applications, the use of strain ratios for material characterization is common (Lankford coefficient, see Section 6.3). Therefore, the ratios of width-strain to thickness-strain (r-value, Lankford coefficient) need to be converted for the use in the Abaqus Hill model (Simulia, 2012).

With the rolling direction aligned with the 11-direction, the r-value in rolling direction reads

$$r_0 = \frac{\mathrm{d}\varepsilon_{22}}{\mathrm{d}\varepsilon_{33}} = \frac{H}{G}\,, \tag{B.12}$$

and the r-value in transverse direction is defined by

$$r_{90} = \frac{\mathrm{d}\varepsilon_{11}}{\mathrm{d}\varepsilon_{33}} = \frac{H}{F}\,. \tag{B.13}$$

The r-value at angle α is described in Simulia (2012) by

$$r_\alpha = \frac{H + (2N - F - G - 4H)\sin^2\alpha\cos^2\alpha}{F\sin^2\alpha + G\cos^2\alpha} \,. \tag{B.14}$$

For sheet metals, the following relations can be derived for the R_{ij} values that need to be specified in the Abaqus Hill model (see Simulia, 2012):

$$R_{11} = 1 \,, \tag{B.15}$$

$$R_{22} = \sqrt{\frac{r_0(1+r_{90})}{r_{90}(1+r_0)}} \,, \tag{B.16}$$

$$R_{33} = \sqrt{\frac{r_{90}(1+r_0)}{r_0+r_{90}}} \,, \tag{B.17}$$

$$R_{12} = \sqrt{\frac{3(1+r_0)r_{90}}{(2r_{45}+1)(r_0+r_{90})}} \,, \tag{B.18}$$

$$R_{13} = 1 \,, \tag{B.19}$$

$$R_{23} = 1 \,. \tag{B.20}$$

For each r-value r_0, r_{45} and r_{90}, the arithmetic mean is taken from the experimentally measured Lankford coefficients in Fig. 6.4. Thereby, the parameters R_{ij} in eq. (B.15) – (B.20) for the aluminum alloy EN AW-3104 can be calculated and they are summarized in Tab. B.1.

Table B.1: Parameters R_{ij} for the Hill yield surface

R_{11}	R_{22}	R_{33}	R_{12}	R_{13}	R_{23}
1	1.39173	1.03655	1.0528	1	1

Appendix C

Algorithmic tangent operator: partial derivatives

In order to improve the readability, the superscript $n+1$ is omitted from all quantities at the time step t^{n+1}. If the elastic part of deformation is small, the approximation $\boldsymbol{C}_e \approx \boldsymbol{I}$ holds. With the implicit Euler integration-scheme from eq. (4.11) and eq. (4.12), the partial derivatives of eq. (4.14) lead to:

$$\frac{\partial \boldsymbol{\tau}}{\partial \boldsymbol{F}_e} \approx \frac{1}{2}(\boldsymbol{F}_e \star \mathbb{C}\left[\boldsymbol{I} \square \boldsymbol{F}_e^{\mathsf{T}} + (\boldsymbol{F}_e \square \boldsymbol{I})^{\mathsf{T_R}}\right] , \tag{C.1}$$

$$\frac{\partial \boldsymbol{L}}{\partial \boldsymbol{D}} = \mathbb{I}^S , \tag{C.2}$$

$$\frac{\partial \boldsymbol{F}_e}{\partial \boldsymbol{L}} = -\left(\frac{\partial \boldsymbol{G}}{\partial \boldsymbol{F}_e}\right)^{-1} \frac{\partial \boldsymbol{G}}{\partial \boldsymbol{L}} , \tag{C.3}$$

$$\frac{\partial \tau_\alpha}{\partial \boldsymbol{F}_e} \approx \boldsymbol{F}_e \mathbb{C}\left[\boldsymbol{M}_\alpha\right] , \tag{C.4}$$

$$\frac{\partial \boldsymbol{G}}{\partial \boldsymbol{L}} = -\Delta t \boldsymbol{I} \square \boldsymbol{F}_e . \tag{C.5}$$

Substituting these partial derivatives with eq. (4.14), the algorithmic tangent, in eq. (4.13) leads to

$$\mathbb{C}^{\text{alg}} = \frac{1}{2\det(\boldsymbol{F})}(\boldsymbol{F}_e \star \mathbb{C})\left[\boldsymbol{I} \square \boldsymbol{F}_e^{\mathsf{T}} + (\boldsymbol{F}_e \square \boldsymbol{I})^{\mathsf{T_R}}\right]\left(\frac{\partial \boldsymbol{G}}{\partial \boldsymbol{F}_e}\right)^{-1} (\boldsymbol{I} \square \boldsymbol{F}_e)\, \mathbb{I}^S , \tag{C.6}$$

with the partial derivative

$$\frac{\partial \boldsymbol{G}}{\partial \boldsymbol{F}_e} = \mathbb{I} - \Delta t \left(\boldsymbol{L} \square \boldsymbol{I} - \boldsymbol{I} \square \boldsymbol{L}_p - \mathbb{F}_1 \right) , \tag{C.7}$$

and the definition

$$\mathbb{F}_1 = \sum_{\alpha=1}^{N} \frac{m \dot{\gamma}_0}{\tau^D} \left\langle \frac{|\tau_\alpha| - \tau^C}{\tau^D} \right\rangle^{m-1} (\boldsymbol{F}_e \boldsymbol{M}_\alpha) \otimes (\boldsymbol{F}_e \mathbb{C}[\boldsymbol{M}_\alpha]) , \tag{C.8}$$

which is the algorithmic tangent operator from Böhlke et al. (2005).

List of figures

List of tables

Bibliography

Aretz, H., Barlat, F., 2013. New convex yield functions for orthotropic metal plasticity. International Journal of Non-Linear Mechanics 51, 97–111.

Aretz, H., Luce, R., Wolske, M., Kopp, R., Goerdeler, M., Marx, V., Pomana, G., Gottstein, G., 2000. Integration of physically based models into FEM and application in simulation of metal forming processes. Modelling and Simulation in Materials Science and Engineering 8 (6), 881.

Asaro, R. J., 1983. Micromechanics of crystals and polycrystals. Advances in Applied Mechanics 23 (1), 115.

Bacroix, B., Gilormini, P., 1995. Finite-element simulations of earing in polycrystalline materials using a texture-adjusted strain-rate potential. Modelling and Simulation in Materials Science and Engineering 3 (1), 1–21.

Balzani, D., Schröder, J., Brands, D., 2010. FE^2-simulation of microheterogeneous steels based on statistically similar rves. In: IUTAM Symposium on Variational Concepts with Applications to the Mechanics of Materials. Springer.

Banabic, D., 2010. Sheet Metal Forming Processes: Constitutive Modeling and Numerical Simulation. Springer.

Banabic, D., Bunge, H.-J., Pöhlandt, K., Tekkaya, A. E., 2000. Formability of metallic materials. Springer.

Barlat, F., Aretz, H., Yoon, J., Karabin, M., Brem, J., Dick, R., 2005. Linear transfomation-based anisotropic yield functions. International Journal of Plasticity 21 (5), 1009–1039.

Beaudoin, A., Dawson, P., Mathur, K., Kocks, U., Korzekwa, D., 1994. Application of polycrystal plasticity to sheet forming. Computer Methods in Applied Mechanics and Engineering 117 (1), 49–70.

Becker, R., Smelser, R., Panchanadeeswaran, S., 1993. Simulations of earing in aluminum single crystals and polycrystals. Modelling and Simulation in Materials Science and Engineering 1 (2), 203.

Bertram, A., 2008. Elasticity and plasticity of large deformations. Springer.

Betten, J., 2001. Kontinuumsmechanik: Elastisches und inelastisches Verhalten isotroper und anisotroper Stoffe. Springer.

Bishop, J., Hill, R., 1951. A theory of the plastic distortion of a polycrystalline aggregate under combined stresses. The London, Edinburgh, and Dublin Philosophical Magazine and Journal of Science 42 (327), 414–427.

Blade, J., 1967. The cube texture in al and its role in the control of earing. Journal of the Australian Institute of Metals 12 (1), 55–63.

Böhlke, T., 2000. Crystallographic texture evolution and elastic anisotropy: Simulation, modeling, and applications. Ph.D. thesis, Otto-von-Guericke University Magdeburg.

Böhlke, T., Haus, U.-U., Schulze, V., 2006. Crystallographic texture approximation by quadratic programming. Acta Materialia 54 (5), 1359–1368.

Böhlke, T., Risy, G., Bertram, A., 2005. A texture component model for anisotropic polycrystal plasticity. Computational Materials Science 32 (3), 284–293.

Bunge, H., 1982. Texture analysis in materials science. Butterworth-Heinemann.

Bunge, H. J., 1965. Zur Darstellung allgemeiner Texturen. Zeitschrift für Metallkunde 56 (12), 872.

Bunge, H. J., Esling, C., 1982. Quantitative texture analysis. Deutsche Gesellschaft für Metallkunde .

Cho, J.-H., Rollett, A., Oh, K., 2004. Determination of volume fractions of texture components with standard distributions in euler space. Metallurgical and Materials Transactions A 35 (3), 1075–1086.

Chung, K., Lee, S., Barlat, F., Keum, Y., Park, J., 1996. Finite element simulation of sheet forming based on a planar anisotropic strain-rate potential. International Journal of Plasticity 12 (1), 93–115.

Chung, K., Shah, K., 1992. Finite element simulation of sheet metal forming for planar anisotropic metals. International Journal of Plasticity 8 (4), 453–476.

Constellium CRV Test Report 43-11-1-3, 2013. Earing profile measurement. 2013.

Constellium CRV Test Report 49-14-1-5, 2013. Crystallographic texture measurement. 2013.

Crumbach, M., Pomana, G., Wagner, P., Gottstein, G., 2001. A Taylor type deformation texture model considering grain interaction and material properties. Part I – fundamentals. Recrystallization and Grain Growth: Proceedings of the 1. Joint International Conference, 27.-31.8.2001, RWTH Aachen, Germany , 1053–1060.

Delannay, L., 2001. Observation and modelling of grain interactions and grain subdivision in rolled cubic polycrystals. Ph.D. thesis, Katholieke Universiteit Leuven.

DIN EN 573-3, 2013. Aluminium and aluminium alloys - Chemical composition and form of wrought products - Part 3: Chemical composition and form of products.

DIN EN ISO 12004-2, 2008. Metallic materials – Sheet and strip – Determination of forming-limit curves – Part 2: Determination of forming-limit curves in the laboratory.

DIN EN ISO 16808, 2014. Metallic materials – Sheet and strip – Determination of biaxial stress-strain curve by means of bulge test with optical measuring systems.

EN 10002-1, 2001. Metallic materials – Tensile testing – Part 1: Method of test at ambient temperature.

Engler, O., Aegerter, J., 2014. Texture and anisotropy in the Al–Mg alloy AA 5005–Part II: Correlation of texture and anisotropic properties. Materials Science and Engineering: A 618, 663–671.

Engler, O., Hirsch, J., 2007. Polycrystal-plasticity simulation of six and eight ears in deep-drawn aluminum cups. Materials Science and Engineering: A 452, 640–651.

Engler, O., Mertens, N., Van Dam, P., 2011. Texture-based design of a convoluted cut-edge for earing-free beverage cans. Journal of Materials Processing Technology 211 (7), 1278–1284.

Engler, O., Randle, V., 2010. Introduction to texture analysis: Macrotexture, microtexture, and orientation mapping. CRC press.

Eshelby, J. D., 1957. The determination of the elastic field of an ellipsoidal inclusion, and related problems. In: Proceedings of the Royal Society of London A: Mathematical, Physical and Engineering Sciences. Vol. 241. The Royal Society.

Feyel, F., 2003. A multilevel finite element method (fe 2) to describe the response of highly non-linear structures using generalized continua.

Computer Methods in Applied Mechanics and Engineering 192 (28), 3233–3244.

Gao, X., Przybyla, C., Adams, B., 2006. Methodology for recovering and analyzing two-point pair correlation functions in polycrystalline materials. Metallurgical and Materials Transactions A 37 (8), 2379–2387.

Gawad, J., Banabic, D., Van Bael, A., Comsa, D. S., Gologanu, M., Eyckens, P., Van Houtte, P., Roose, D., 2015. An evolving plane stress yield criterion based on crystal plasticity virtual experiments. International Journal of Plasticity 75, 141–169.

Gawad, J., Van Bael, A., Eyckens, P., Samaey, G., Van Houtte, P., Roose, D., 2013. Hierarchical multi-scale modeling of texture induced plastic anisotropy in sheet forming. Computational Materials Science 66, 65–83.

Geers, M., Kouznetsova, V., Brekelmans, W., 2010. Multi-scale computational homogenization: Trends and challenges. Journal of computational and applied mathematics 234 (7), 2175–2182.

Gotoh, M., Ishise, F., 1978. A finite element analysis of rigid-plastic deformation of the flange in a deep-drawing process based on a fourth-degree yield function. International Journal of Mechanical Sciences 20 (7), 423–435.

Gross, D., Seelig, T., 2011. Bruchmechanik – Mit einer Einführung in die Mikromechanik. Springer, Heidelberg.

Grujicic, M., Batchu, S., 2002. Crystal plasticity analysis of earing in deep-drawn OFHC copper cups. Journal of materials science 37 (4), 753–764.

Hashin, Z., Shtrikman, S., 1962. A variational approach to the theory of the elastic behaviour of polycrystals. Journal of the Mechanics and Physics of Solids 10 (4), 343–352.

Haupt, P., 2002. Continuum mechanics and theory of materials. Springer Science & Business Media.

Hielscher, R., Schaeben, H., 2008. A novel pole figure inversion method: Specification of the MTEX algorithm. Journal of Applied Crystallography 41 (6), 1024–1037.

Hill, R., 1948. A theory of the yielding and plastic flow of anisotropic metals. Proceedings of the Royal Society of London 193, 281–297.

Hill, R., 1963. Elastic properties of reinforced solids: Some theoretical principles. Journal of the Mechanics and Physics of Solids 11 (5), 357–372.

Hill, R., 1967. The essential structure of constitutive laws for metal composites and polycrystals. Journal of the Mechanics and Physics of Solids 15 (2), 79–95.

Hill, R., 1990. Constitutive modelling of orthotropic plasticity in sheet metals. Journal of the Mechanics and Physics of Solids 38 (3), 405–417.

Hirsch, J., 2005. Texture evolution and earing in aluminium can sheet. In: Materials Science Forum. Vol. 495. Trans Tech Publications.

Hosford, W., Caddell, R., 2011. Metal Forming: Mechanics and Metallurgy. Cambridge University Press.

Hosford, W., Galdos, A., 1990. Lower bound yield locus calculations. Textures and Microstructures 12, 89.

Hutchinson, W., Oscarsson, A., Karlsson, A., 1989. Control of microstructure and earing behaviour in aluminium alloy AA 3004 hot bands. Materials Science and Technology 5 (11), 1118–1127.

Inal, K., Wu, P., Neale, K., 2000. Simulation of earing in textured aluminum sheets. International Journal of Plasticity 16 (6), 635–648.

Jain, M., Lloyd, D., MacEwen, S., 1996. Hardening laws, surface roughness and biaxial tensile limit strains of sheet aluminium alloys. International Journal of Mechanical Sciences 38 (2), 219–232.

Jöchen, K., 2013. Homogenization of the linear and non-linear mechanical behavior of polycrystals. Ph.D. thesis, Karlsruhe Institute of Technology (KIT).

Jöchen, K., Böhlke, T., 2012. Prediction of texture evolution in rolled sheet metals by using homogenization schemes. Key Engineering Materials 504–506, 649–654.

Jöchen, K., Böhlke, T., 2013. Representative reduction of crystallographic orientation data. Journal of Applied Crystallography 46 (4), 960–971.

Kalidindi, S. R., Bronkhorst, C. A., Anand, L., 1992. Crystallographic texture evolution in bulk deformation processing of FCC metals. Journal of the Mechanics and Physics of Solids 40 (3), 537–569.

Kanouté, P., Boso, D., Chaboche, J., Schrefler, B., 2009. Multiscale methods for composites: A review. Archives of Computational Methods in Engineering 16 (1), 31–75.

Kocks, U., Chandra, H., 1982. Slip geometry in partially constrained deformation. Acta Metallurgica 30 (3), 695–709.

Kocks, U., Tomé, C., Wenk, H., 2000. Texture and Anisotropy: Preferred Orientations in Polycrystals and Their Effect on Materials Properties. Cambridge University Press.

Kocks, U., Tome, C., Wenk, H., et al., 1998. Texture and anisotropy. preferred orientations in polycrystals and their effect on material properties. Cambridge University, Cambridge .

Kuramae, H., Ikeya, Y., Sakamoto, H., Morimoto, H., Nakamachi, E., 2010. Multi-scale parallel finite element analyses of LDH sheet formability tests based on crystallographic homogenization method. International Journal of Mechanical Sciences 52 (2), 183–197.

Lankford, W. T., Snyder, S. C., Bausher, J. A., 1950. New criteria for predicting the press performance of deep drawing sheets. Transactions of the American Society of Metals 42, 1197–1205.

Lebensohn, R., Tomé, C., 1993. A self-consistent anisotropic approach for the simulation of plastic deformation and texture development of polycrystals: Application to zirconium alloys. Acta Metallurgica et Materialia 41 (9), 2611–2624.

Leffers, T., 1979. A modified Sachs approach to the plastic deformation of polycrystals as a realistic alternative to the Taylor model. In: 5th International Conference on Strength of Metals and Alloys.

Li, S., Hoferlin, E., Van Bael, A., Van Houtte, P., 2001. Application of a texture-based plastic potential in earing prediction of an IF steel. Advanced engineering materials 3 (12), 990–994.

Mainprice, D., Hielscher, R., Schaeben, H., 2011. Calculating anisotropic physical properties from texture data using the MTEX open-source package. Geological Society, London, Special Publications 360 (1), 175–192.

Mathur, K. K., Dawson, P. R., 1989. On modeling the development of crystallographic texture in bulk forming processes. International Journal of Plasticity 5 (1), 67–94.

Mathur, K. K., Dawson, P. R., Kocks, U., 1990. On modeling anisotropy in deformation processes involving textured polycrystals with distorted grain shape. Mechanics of Materials 10 (3), 183–202.

Melchior, M. A., Delannay, L., 2006. A texture discretization technique adapted to polycrystalline aggregates with non-uniform grain size. Computational materials science 37 (4), 557–564.

Méric, L., Cailletaud, G., Gaspérini, M., 1994. FE calculations of copper bicrystal specimens submitted to tension-compression tests. Acta Metallurgica et Materialia 42 (3), 921–935.

Miehe, C., Rosato, D., Frankenreiter, I., 2010. Fast estimates of evolving orientation microstructures in textured bcc polycrystals at finite plastic strains. Acta Materialia 58 (15), 4911–4922.

Miehe, C., Schröder, J., Schotte, J., 1999. Computational homogenization analysis in finite plasticity simulation of texture development in polycrystalline materials. Computer Methods in Applied Mechanics and Engineering 171 (3), 387–418.

Mura, T., 1987. Micromechanics of Defects in Solids. Kluwer Academic Publishers.

Murnaghan, F. D., 1962. The unitary and rotation groups. Spartan Books.

Naess, S., 1991. Development of earing and texture during temper rolling of the aluminium alloy AA3005 and AA5050. Zeitschrift für Metallkunde 82 (4), 259–264.

Nakamachi, E., Tam, N., Morimoto, H., 2007. Multi-scale finite element analyses of sheet metals by using sem-ebsd measured crystallographic rve models. International Journal of Plasticity 23 (3), 450–489.

Nemat-Nasser, S., 1999. Averaging theorems in finite deformation plasticity. Mechanics of Materials 31 (8), 493–523.

Nemat-Nasser, S., Hori, M., 1999. Micromechanics: Overall properties of heterogeneous materials. Elsevier.

Novák, J., Kaczmarczyk, Ł., Grassl, P., Zeman, J., Pearce, C. J., 2012. A micromechanics-enhanced finite element formulation for modelling heterogeneous materials. Computer Methods in Applied Mechanics and Engineering 201, 53–64.

Olsen, T., 1920. Machines for ductility testing. Proceedings of the American Society for Testing and Materials 20, 398–403.

Park, D., Morris, J., 1993. The Portevin-Le Chatelier effect in an Al-Mn-Mg alloy. Scripta Metallurgica et Materialia 29 (3), 365 – 369.

Pedersen, O., Leffers, T., 1987. Modelling of Plastic Heterogeneity in Deformation of Single-Phase Materials. Risø National Laboratory, pp. 147–172.

Phan Van, T., Jöchen, K., Böhlke, T., 2012. Simulation of sheet metal forming incorporating EBSD data. Journal of Materials Processing Technology 212 (12), 2659–2668.

Ponte Castañeda, P., Suquet, P., 1998. Nonlinear composites. Advances in Applied Mechanics 34, 171–302.

Raabe, D., Roters, F., 2004. Using texture components in crystal plasticity finite element simulations. International Journal of Plasticity 20 (3), 339–361.

Ren, B., Das, S., 1998. Aluminum alloys for packaging III. TMS (Warrendale, PA) , 49.

Renard, J., Marmonier, M., 1987. Etude de l'initiation de l'endommagement dans la matrice d'un matériau composite par une méthode d'homogénéisation. La Recherche aérospatiale (6), 43–51.

Reuss, A., 1929. Berechnung der Fließgrenze von Mischkristallen auf Grund der Plastizitätsbedingung für Einkristalle. ZAMM-Journal of Applied Mathematics and Mechanics/Zeitschrift für Angewandte Mathematik und Mechanik 9 (1), 49–58.

Roters, F., 2011. Advanced material models for the crystal plasticity finite element method: development of a general CPFEM framework. Habilitation.

Roters, F., Eisenlohr, P., Hantcherli, L., Tjahjanto, D., Bieler, T., Raabe, D., 2010. Overview of constitutive laws, kinematics, homogenization and multiscale methods in crystal plasticity finite-element modeling: Theory, experiments, applications. Acta Materialia 58 (4), 1152–1211.

Sachs, G., 1928. Zur Ableitung einer Fließbedingung. Zeitschrift des Vereines Deutscher Ingenieure 72, 734–736.

Schaeben, H., 1997. A simple standard orientation density function: The hyperspherical de La Vallée Poussin kernel. Physica Status Solidi (B) 200 (2), 367–376.

Schaeben, H., Hielscher, R., 2016. MTEX: Matlab toolbox for analyzing and modeling crystallographic textures by means of EBSD or pole figure data. http://mtex-toolbox.github.io/, visited on January 6th, 2016.

Schulze, V., Bertram, A., Böhlke, T., Krawietz, A., 2009. Texture-based modeling of sheet metal forming and springback. Technische Mechanik 29 (2), 135–159.

Schwartz, A. J., Kumar, M., Adams, B. L., Field, D. P., 2009. Electron backscatter diffraction in materials science. Vol. 2. Springer.

Segurado, J., Lebensohn, R. A., LLorca, J., Tomé, C. N., 2012. Multiscale modeling of plasticity based on embedding the viscoplastic self-consistent formulation in implicit finite elements. International Journal of Plasticity 28 (1), 124–140.

Simo, J., Hughes, T. J., 1998. Computational Inelasticity. Springer-Verlag, Berlin.

Simulia, 2012. ABAQUS 6.12 Manual. 3DS Dassault Systèmes, http://uni-license1.uni-karlsruhe.de:2080/v6.12/index.html, visited on October 12th, 2015.

Suisse Technology Partners AG Test Report E-13-025, 2013. Hydraulic bulge test according to ISO/CD 16808. 2013.

Suisse Technology Partners AG Test Report E-13-030, 2013. Nakajima test according to ISO 12004-2. 2013.

Suisse Technology Partners AG Test Report M-13-005, 2013. Tensile test according to EN 10002-1. 2013.

Taylor, G., 1938. Plastic strain in metals. Journal of the Institute of Metals 62, 307–324.

Tomé, C., Maudlin, P., Lebensohn, R., Kaschner, G., 2001. Mechanical response of zirconium–I. derivation of a polycrystal constitutive law and finite element analysis. Acta Materialia 49 (15), 3085–3096.

Torquato, S., 2002. Random Heterogeneous Materials: Microstructure and Macroscopic Properties. Springer.

Tóth, L. S., Van Houtte, P., 1992. Discretization techniques for orientation distribution functions. Textures and Microstructures 19 (4), 229–244.

Van Houtte, P., 1982. On the equivalence of the relaxed Taylor theory and the Bishop-Hill theory for partially constrained plastic deformation of crystals. Materials Science and Engineering 55 (1), 69–77.

Van Houtte, P., Delannay, L., Samajdar, I., 1999. Quantitative prediction of cold rolling textures in low-carbon steel by means of the Lamel model. Texture, Stress, and Microstructure 31 (3), 109–149.

Van Houtte, P., Gawad, J., Eyckens, P., Van Bael, B., Samaey, G., Roose, D., 2012. Multi-scale modelling of the development of heterogeneous distributions of stress, strain, deformation texture and anisotropy in sheet metal forming. Procedia IUTAM 3, 67–75.

Van Houtte, P., Li, S., Seefeldt, M., Delannay, L., 2005. Deformation texture prediction: From the Taylor model to the advanced Lamel model. International Journal of Plasticity 21 (3), 589–624.

Voce, E., 1955. A practical strain-hardening function. Metallurgia 51 (307), 219–226.

Voigt, W., 1889. Über die Beziehung zwischen den beiden Elastizitätskonstanten isotroper Körper. Annalen der Physik 274 (12), 573–587.

von Mises, R., 1928. Mechanik der plastischen Formänderung von Kristallen. Zeitschrift für Angewandte Mathematik und Mechanik (ZAMM) 8 (3), 161–185.

Walde, T., Riedel, H., 2007. Simulation of earing during deep drawing of magnesium alloy AZ31. Acta Materialia 55 (3), 867–874.

Westerman, E., 1993. Silicon: A vital alloying element in aluminum beverage can body stock. Aluminum Alloys for Packaging , 1–16.

Willis, J., 1977. Bounds and self-consistent estimates for the overall of anisotropic composites. Journal of the Mechanics and Physics of Solids 25, 185–202.

Willis, J., 1981. Variational and related methods of the overall properties of composites. Advances in Applied Mechanics 21, 1–78.

Wriggers, P., 2008. Nonlinear finite element methods. Springer Science & Business Media.

Yoon, J., Barlat, F., Chung, K., Pourboghrat, F., Yang, D., 2000. Earing predictions based on asymmetric nonquadratic yield function. International Journal of Plasticity 16 (9), 1075–1104.

Yoon, J., Barlat, F., Dick, R., Karabin, M., 2006. Prediction of six or eight ears in a drawn cup based on a new anisotropic yield function. International Journal of Plasticity 22 (1), 174–193.

Zhou, Y., Jonas, J., Savoie, J., Makinde, A., MacEwen, S., 1998. Effect of texture on earing in FCC metals: Finite element simulations. International Journal of Plasticity 14 (1), 117–138.

Schriftenreihe Kontinuumsmechanik im Maschinenbau Karlsruher Institut für Technologie (KIT)

(ISSN 2192-693X)

Herausgeber: Prof. Dr.-Ing. Thomas Böhlke

Die Bände sind unter www.ksp.kit.edu als PDF frei verfügbar oder als Druckausgabe bestellbar.

Band 1 Felix Fritzen
Microstructural modeling and computational homogenization of the physically linear and nonlinear constitutive behavior of micro-heterogeneous materials. 2011
ISBN 978-3-86644-699-1

Band 2 Rumena Tsotsova
Texturbasierte Modellierung anisotroper Fließpotentiale. 2012
ISBN 978-3-86644-764-6

Band 3 Johannes Wippler
Micromechanical finite element simulations of crack propagation in silicon nitride. 2012
ISBN 978-3-86644-818-6

Band 4 Katja Jöchen
Homogenization of the linear and non-linear mechanical behavior of polycrystals. 2013
ISBN 978-3-86644-971-8

Band 5 Stephan Wulfinghoff
Numerically Efficient Gradient Crystal Plasticity with a Grain Boundary Yield Criterion and Dislocation-based Work-Hardening. 2014
ISBN 978-3-7315-0245-6

Band 6 Viktor Müller
Micromechanical modeling of short-fiber reinforced composites. 2016
ISBN 978-3-7315-0454-2

Schriftenreihe Kontinuumsmechanik im Maschinenbau
Karlsruher Institut für Technologie (KIT)
(ISSN 2192-693X)

Band 7 Florian Rieger
Work-hardening of dual-phase steel. 2016
ISBN 978-3-7315-0513-6

Band 8 Vedran Glavas
Micromechanical Modeling and Simulation of Forming Processes. 2017
ISBN 978-3-7315-0602-7